Thomas Scott Dean, Ph.D.
Director of Programs in Architectural Engineering
University of Kansas

THERMAL STORAGE

© 1978
THE FRANKLIN INSTITUTE PRESS
Philadelphia, Pennsylvania.

All rights reserved. No part of this book may be reproduced in any form, for any purpose or by any means, abstracted, or entered into any data base, electronic or otherwise, without specific permission in writing from the publisher.

Current printing (last digit):
5 4 3 2
THE FRANKLIN INSTITUTE PRESS NUMBER: D/1078
ISBN Number: 0-89168-005-5
Library of Congress Catalog Card Number: 78-3211
Printed in the United States of America.

Also in the Solar Energy Series:

Solar Collector Design
  D. K. Edwards

Solar Cooling
  Paul J. Wilbur and Susumu Karaki

Bioconversion: Fuels From Biomass
  E. E. Robertson

Wind Energy
  B. Wolff and H. Meyer

High Temperature Thermal Storage
  R. Turner

Winds and Wind System Performance
  C. G. Justus

TABLE OF CONTENTS

I. Introduction 9

II. Passive and Active Heating Systems 11

III. Storage Materials and Construction of Container 15
    Sensible Heat 15
    Losses from Storage 18
    Heat Conduction in Storage 19
    Water 20
    Rock 25
    Hybrid Systems 29
    Performance of Thermal Storage Systems 31
    Heat of Fusion 31

IV. Pioneers in Thermal Storage 36
    Trombe Houses 37
    David Wright House 38
    Atascadero House 38
    Baer House 40
    Dean House 40
    Performance Verification for Passive Buildings 43

V. Thermal Storage Strategies 48
    Thermal Storage for Domestic Hot Water 53
    Other Storage Strategies 58

    References 59

    Appendix (Table) 61

LIST OF FIGURES

II-1. A Passive Building 11
II-2. An Active Solar Energy System 12
II-3. Water Storage Tank for Zero Energy House 14
III-1. Simulated Two-Section Storage Tank 21
III-2. Stratified Storage Tank 22
III-3. a. Instrumented Storage Tank 23
      b. Temperature Profile for Instrumented Tank 24
III-4. a. Horizontal Pebble Bed 26
      b. Vertical Pebble Bed 27
III-5. Temperature Profile for a Vertical Pebble Bed 28
III-6. Hybrid Air and Water Storage 30
III-7. Water-Filled Containers on Racks Used with Air Systems 31
III-8. Comparison of Water and Phase-Changing Glauber's Salts 34
IV-1. Atascadero House, Roof Section 39
IV-2. Section of Dean House 41
IV-3. a. Effect of Storage Mass 42
      b. Effect of Glass Area 43
IV-4. a. Time Response for Six-Inch Masonry Wall for
        One Week Period 44
      b. Time Response for Twelve-Inch Masonry Wall for
        One Week Period 45
      c. Time Response for Twenty-four Inch Masonry Wall for
        One Week Period 46
V-1. A Solar Pond Collector 48
V-2. Solar Assist for a Water-to-air Heat Pump 52
V-3. Solar Assist for an Air-to-air Heat Pump 53
V-4. Thermosyphonic Domestic Water Heating Schematic 55
V-5. Schematic for Solar Assist to Conventional Water Heater 56
V-6. Modification of Figure V-5 for
     Minimizing Conventional Energy Input 57
V-7. Schematic for Efficient Solar Heat Pre-Heat System 58

LIST OF TABLES

I. Properties of Solids 16
II. Sensible Heat Storage of Common Materials 17
III. Properties of Salt Hydrates 33
IV. Performance of a Commercially Available Three-Ton Heat Pump 50
V. Hot Water Demand Characteristics 54

Appendix
Conversion Factors: Common Relationships 61

1. INTRODUCTION

It is not difficult to imagine that with the discovery of fire our prehistoric ancestors realized rather sooner than later that somehow this heat energy was locked up in the wood, leaves, grasses, and other plant material used as their fuel. This energy, as with all other energy used on this planet (with the exception of nuclear energy), had its origin in the million degree fusion reactor we call our sun. The sun's "cool" 6000°K photosphere emits this energy in the form of electromagnetic radiation, some of which, after travelling over 90,000,000 miles through space, is absorbed by the green leaves of growing plants. Through the process of photosynthesis this energy is converted into cellulose and lignin, the building blocks of plant material, ready to be converted to heat energy upon ignition. That these plant materials, particularly wood, are acceptable thermal storage devices is adequately demonstrated by the fact that wood was humanity's *primary* heat source until about 200 years ago.[1]

Some of these plant materials, instead of decaying on the forest floor or open prairie, were converted into coal. Since the moisture and air spaces of the plants had been largely eliminated, the resulting material possessed considerably more heat content than wood on either a volume or weight basis. This economy of thermal storage resulted in relatively easier collection, transportation, and utilization than wood offered. During the last century, large coal-powered oceangoing ships engaged in commerce, travel, and war. Coal-fired locomotives thundered along twin steel rails, effectively uniting all parts of this country. Coal fueled the blast furnaces and engines of industry which made possible both locomotives and rails. Coal was burned to produce the steam necessary for a fledgling electric industry. By the turn of this century, coal had largely replaced wood as a primary fuel except for some home heating or a frontier outpost.

About 100 years ago, petroleum was discovered in Pennsylvania; it was later found in Texas and in many parts of the world. The first wells were shallow and often the deposit was accompanied by natural gas. This gas was usually burned at the well head as an undesirable by-product. Just as coal was easier to collect, transport, and utilize than wood, so petroleum and later natural gas claimed the same advantages over coal. How much simpler to sink a pipe into the reservoir than to send miners into the coal seam. With the completion of the Big Inch pipeline during World War II, the escalating cost of coal brought on by apparent intransigence on the part of both miners and operators, and the relative ease of burning petroleum and natural gas, the demise of coal as the primary fuel was assured.

We are tempted to think of electricity as our primary energy source today, but this concept is valid only for the end user. Some electric power is generated by nuclear reactors which use the atom as energy storage. Some electric power is generated by hydroelectric plants which utilize elevated water reservoirs as energy storage. But the bulk of our electric energy is produced in steam generating plants driven by burning fossil fuel.

Even if it is not necessary to consider wood, coal, petroleum, and natural gas as thermal storage materials it seems desirable to do so. For in so doing, it becomes apparent that our societies are rapidly eliminating their most convenient, efficient, and economical energy storage. Only one of these, wood, is renewable in a lifetime. It may be that humanity faces a greater challenge in devising methods for storing energy than in harvesting it. If only our scientific establishment could discover a way to convert electricity (or solar radiation) into petroleum!

The energy crisis of the mid-1960's and one of its consequences, the oil embargo of 1973, caused some people to question the wisdom of a single energy source.

During the last decade the United States has taken a few tentative steps in developing alternate forms of energy such as solar, wind, tidal, wave, and geothermal. Although these steps are far less forceful than many of us would wish, still this country leads the world in its commitment to the development of alternate energy sources. This leadership is due in no small part to earlier pioneering investigations carried on by Maria Telkes, John Yelliott, Eric Farber, George Löf, and other daring souls. More recently, they have been joined by a plethora of scientific laboratories and radical counterculturists operating outside the academies. The former group may be commended for its attention to the measurement of the properties of solar collection and storage systems. The second group should be commended no less for its creativity and resourcefulness as related to these systems. Finally, both groups should be commended for bridging the philosophical gulf separating them in order to contribute to society as a whole.

Among the efforts to develop alternate energy sources, flat plate solar technology appropriate for space and water heating and space cooling has progressed beyond the experimental stage. In many cases, it is now cost-effective to install such systems when comparison is made with electric energy. One can only assume that solar energy systems will become more attractive to mainstream Americans as the cost of conventional energy sources continues to escalate.

However, mainstream America has its own notions about energy system performance—about desired temperature and quantity of domestic hot water, and about the temperature variation in homes, offices, and schools. It is at this point that thermal storage systems become a pivotal issue in the design of a solar energy system. It is to this issue that the present work is dedicated.

2. PASSIVE AND ACTIVE HEATING SYSTEMS

Solar heating systems may be classified as *passive* or *active*. A passive or direct gain system is one is which the structure itself, through the strategic placement of glazing, serves as the collector. The insulated thermal mass of the structure acts as thermal energy storage. Such systems may be characterized as simple and inexpensive, free from equipment malfunction, and requiring zero, or almost zero, conventional energy input. However, these systems will not always meet the performance specifications expected by a wide spectrum of people. Inside temperatures often vary by 20°F or more during a 24-hour period. Adequate physical control of incoming short wave radiation and outgoing heat through glazing is desirable. Unless some form of forced circulation of air is employed, some spaces may be uncomfortably warm, while others colder than desired (Figure 2-1).

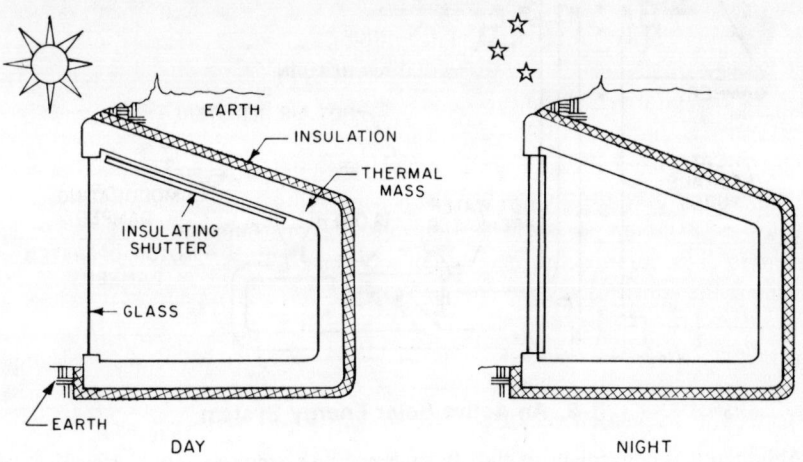

II-1. A Passive Building

Typically, the thermal storage system of a passive building consists of the materials used in its construction, although sometimes additional mass is included. These materials are dense in order to secure the greatest mass. Since large quantities may be used, the cost per unit weight becomes a constraint. Common building materials used for this purpose include concrete, brick, stone, earth, and adobe. On a weight basis, wood offers about the same thermal storage as earthy material. Several passive buildings employ additional thermal storage in the form of water-filled containers such as steel drums or indoor swimming pools.

An active solar system for space or water heating consists of an intentionally designed collector array which harvests heat energy. Heat is removed from the collectors by air or liquid. Conventionally, plain water or antifreeze solution is the heat transfer liquid, although some proprietory materials are available. Usually these active systems are capable of directly delivering heat to the interior spaces. More often than not, however, the period of maximum heat collection is different from the period of maximum space heat demand, so that separate thermal energy storage is needed. In order to be usable, this energy must be at a temperature higher than the design temperature of the space to be heated. The magnitude of this temperature elevation is a function of the type of heating system, transport medium, and number and character of any heat exchangers (Figure 2-2).

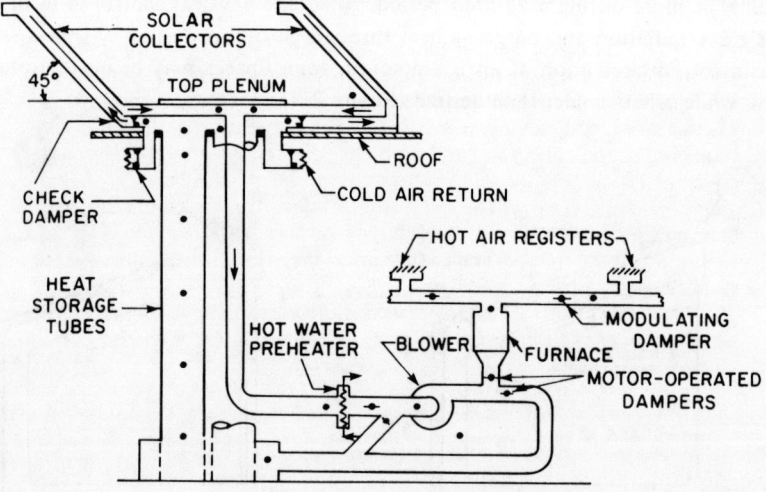

II-2. An Active Solar Energy System

Although it is convenient to classify solar energy systems as active or passive, the fact is that all new building should be subject to passive design and existing buildings should be modified in the passive mode insofar as practicable. As indicated above, it is unlikely that in most climates a passive design alone will meet the desired performance specifications of most people. Furthermore, although passive designs enjoy a favored position among those solar designers seeking simplification and economy, precise and predictable design of such buildings requires a knowledge of radiation phenomena, heat transfer, and heat storage far beyond the capabilities of most architects and engineers. However, the recognition of and appropriate response to passive design will reduce to a minimum the size of collector arrays, required thermal storage system, system circuitry, and the energy from conventional sources necessary to drive the solar supplement.

In order to supply energy for domestic water heating or space heating, minimum temperature of the thermal storage system generally must be above 100°F, but seldom needs to exceed 160°F. When the storage system is required to supply heat to existing absorption chillers, the temperature range is 160°F–200°F, but the lower temperatures result in reduced cooling capacity. As will be explained later, the minimum temperature for space heating may be lowered significantly when this energy is used to supplement a heat pump. Obviously, the temperature of the thermal storage system at the end of a period of energy harvest must be higher than the specified minimum.

Determination of the optimum storage capacity for any given installation is a function of at least three considerations. These are:

1. *Effect on collector performance.* A smaller storage unit must be maintained at a higher mean temperature. Consequently, its maximum temperature at the end of a collection period may be unacceptably high. Furthermore, the temperature difference between storage (and fluid returning to the collectors) and ambient is increased, thereby decreasing collector efficiency.
2. *Thermal losses.* Heat will be lost from storage not only as a function of the external surface area and conductivity of the container, but also as a function of the temperature difference between storage and its surroundings.
3. *Economics.* As with all energy systems, the cost of the storage material, its container, the space it occupies, its maintenance, and the cost of storing and retrieving the energy must be considered.

Based on records from existing systems and numerous computer models, it appears that storage adequate for periods from a few hours to a few days is most economical, although some experimental long-term storage systems, such as Denmark's "Zero Energy House" and Canada's "Provident House," are in operation.

The Zero Energy House was designed and constructed by the Technical University of Denmark during the spring of 1975.[2] A single family, one-story dwelling, it consists of two 60 m^2 living spaces separated by a 70 m^2 unheated atrium. A vertical south-facing liquid-cooled collector array is mounted on the upper wall of the atrium. Heat is transferred to a buried 30 m^3 steel storage tank external to the building. Despite 60 cm (24 inches) of mineral wool surrounding the tank and the absence of ground water, 40 percent of the stored energy is accounted for by losses. Had the tank been buried beneath the house, this loss would have been much less and there would have been an uncontrolled heat gain for the building (Figure 2-3).

Canada's Provident House, sponsored by the federal and Ontario governments, is designed to be totally solar heated. Unlike the Denmark house, storage for this system is located beneath the building and has a capacity of 63,000 gallons. No long-term performance data are yet available.

An additional long-term storage concept is being investigated at Oak Ridge, Tennessee. This system, the Annual Cycle Energy System (ACES), employs a heat pump in conjunction with a very large water (or ice) storage reservoir. The heat pump operates by extracting heat from, or rejecting heat to, the storage reservoir on a seasonal basis. The idea is ingenious and from a technological viewpoint presents no insurmountable obstacles. However, it is unclear at this time if the ACES system will prove economically viable in small-scale applications.

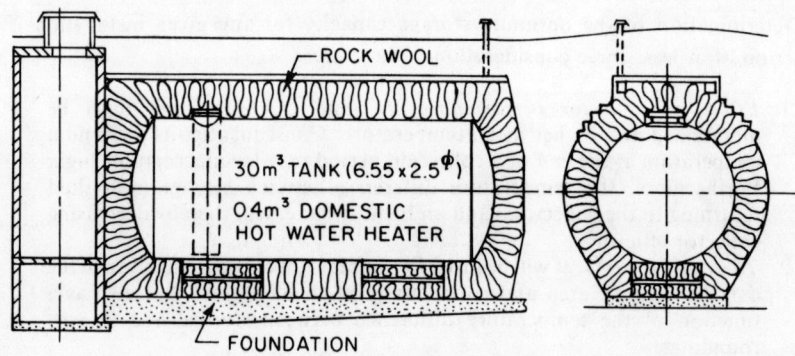

II-3. Water Storage Tank for Zero Energy House

3. STORAGE MATERIALS AND CONSTRUCTION OF CONTAINER

Energy may be stored as a sensible heat in either a liquid or solid medium (or in some cases a combination of the two), or as heat of fusion in selected salts. Of these two methods, storage as sensible heat is best understood and most reliable at this time. It may be regarded as a mature technology. Storage as heat of fusion (latent heat) is not understood as well. However, as additional research is applied to this important area, one may expect an expanded use of thermal storage systems utilizing latent heat.

Sensible Heat. Sensible heat is that heat which may be measured by an ordinary wall thermometer. The specific heat, C_p, of a substance is the amount of heat necessary to raise the temperature of a unit mass one degree. By definition, the specific heat of water is 1.0, that is, for water $C_p = 1.0$ BTU/lb-°F when English units are used. Specific heat values for common materials are given in Tables 1 and 2.

The amount of available energy which can be stored as sensible heat in a thermal storage system is

$$H = mC_p \, \Delta T \tag{3-1}$$

where

H = heat capacity of system

m = mass of storage medium

C_p = specific heat of storage medium

ΔT = temperature limits between which the system operates (maximum temperature of store ⩾ lowest temperature at which usable energy can be extracted).

Example 1: Determine the storage capactiy of a 5 ft × 8 ft × 4.5 ft rock bin filled with limestone rock (C_p = .217 BTU/lb -°F) and containing 20 percent voids. Maximum temperature in storage is 160°F and air off the rocks must be at least 93°F. Limestone weight 155 lb /ft^3.

$H = 155$ lb/ft^3 × (5 ft × 8 ft × 4.5 ft) × .80 × .217 BTU/lb-°F × (160° − 93°)
$= .325 × 10^6$ BTU

Example 2: Determine the storage capacity of a 5 ft × 8 ft × 4.5 ft water tank if the maximum tank temperature is 160°F and water in an internal heat exchanger must be at least 93°F. Approach temperature for the heat exchanger is 6°F.

Table 1. Properties of Solids
(Values are for room temperature unless otherwise noted in brackets)

Material Description	Specific Heat BTU/lb-°F	Density lb/ft^3
Aluminum (alloy 1100)	0.214	171
Aluminum Bronze (76% Cu, 22% Zn, 2% Al)	0.09	517
Alundum (aluminum oxide)	0.186	
Asbestos: fiber	0.25	150
insulation	0.20	36
Ashes, wood	0.20	40
Asphalt	0.22	132
Bakelite	0.35	81
Bell metal	0.086 [122]	
Bismuth tin	0.040	
Brick, building	0.2	123
Brass:		
red (85% Cu, 15% Zn)	0.09	548
yellow (65% Cu, 35% Zn)	0.09	519
Bronze	0.104	530
Cadmium	0.055	540
Carbon (gas retort)	0.17	
Cardboard		
Cellulose	0.32	4.4
Cement (Portland clinker)	0.16	120
Chalk	0.215	143
Charcoal (wood)	0.20	15
Chrome Brick	0.17	200
Clay	0.22	63
Coal	0.3	90
Coal Tars	0.35 [104]	75
Coke (petroleum, powdered)	0.36 [752]	62
Concrete (stone)	0.156 [392]	144
Copper (electrolytic)	0.092	556
Cork (granulated)	0.485	5.4
Cotton (fiber)	0.319	95
Cryolite (AIF₃3NaF)	0.253	181
Diamond	0.147	151
Earth (dry and packed)		95
Felt		20.6
Fireclay brick	0.198 [212]	112
Flourspar (CaF₂)	0.21	199
German Silver (nickel silver)	0.09	545
Glass:		
crown (soda-lime)	0.18	154
flint (lead)	0.117	267
pyrex	0.20	139
"wool"	0.157	3.25
Gold	0.0312	1208
Graphite:		
powder	0.165	
"Karbate" (impervious)	0.16	117
Gypsum	0.259	78
Hemp (fiber)	0.323	93
Ice: [32°F]	0.487	57.5
[-4°F]	0.465	
Iron:		
cast	0.12 [212]	450
wrought		485
Lead	0.0309	707
Leather (sole)		62.4
Limestone	0.217	103
Linen		

Material Description	Specific Heat BTU/lb-°F	Density lb/ft^3
Litharge (lead monoxide)	0.055	490
Magnesia:		
powdered	0.234 [212]	49.7
light carbonate		13
Magnesite brick	0.222 [212]	158
Magnesium	0.241	108
Marble	0.21	162
Nickel	0.105	555
Paints:		
White lacquer		
White enamel		
Black lacquer		
Black shellac		63
Flat black lacquer		
Aluminum lacquer		
Paper	0.32	58
Paraffin	0.69	56
Plaster		132
Platinum	0.032	1340
Porcelain	0.18	162
Pyrites (Copper)	0.131	262
Pyrites (Iron)	0.136 [156]	310
Rock Salt	0.219	136
Rubber:		
Vulcanized (soft)	0.48	68.6
(hard)		74.3
Sand	0.191	94.6
Sawdust		12
Silica	0.316	140
Silver	0.0560	654
Snow (freshly fallen)		7
(at 32°F)		31
Steel (mild)	0.12	489
Stone (quarried)	0.2	95
Tar:		
pitch	0.59	67
bituminous		75
Tin	0.0556	455
Tungsten	0.032	1210
Wood:		
Hardwoods:	0.45/0.65	23/70
Ash, white		43
Elm, American		36
Hickory		50
Mahogany		34
Maple, sugar		45
Oak, white	0.570	47
Walnut, black		39
Softwoods:		22/46
Fir, white	0.65	27
Pine, white	0.67	27
Spruce		26
Wool:		
Fiber	0.325	82
Fabric		6.9/20.6
Zinc:		
Cast	0.092	445
Hot-rolled	0.094	445
Galvanizing		

(From Anderson, p. 349)

Storage Materials and Construction of Container

Table 2. Sensible Heat Storage of Common Materials—Condensed

Sensible-Heat Storage	Specific Heat BTU/lb-°F	True Density, lb/ft³	Heat Capacity, BTU/ft³-°F	
			No Voids	30% Voids
Water	1.00	62	62	43
Scrap Iron	0.12	490	59	41
Magnetite (Fe$_3$O$_4$)	0.18	320	57	40
Scrap Aluminum	0.23	170	39	27
Concrete	0.27	140±	38	26
Stone	0.21	170±	36	25
Brick	0.20	140	28	20
Sodium (to 208 F)	0.23	59	14	—

Figure V·D·1. Heat storage materials. Source: (HOT.)

(From Anderson, p. 216)

$$H = 62.4 \text{ lb/ft}^3 \times (5 \text{ ft} \times 8 \text{ ft} \times 4.5 \text{ ft}) \times 1.0 \text{ BTU/lb-°F} \times [160° - (93° + 6°)]$$
$$= .685 \times 10^6 \text{ BTU}$$

Example 3: Determine the volume of limestone necessary to store the heat that can be stored by a given volume of water. Assume 30 percent voids in the rock for air passage.

Solution: Assume one cubic foot of water and a 1°F temperature difference. Then

$$H = 1 \text{ ft}^3 \times 62.4 \text{ lb/ft}^3 \times 1.0 \text{ BTU/lb-°F} = 62.4 \text{ BTU}$$

For rock

$$H = 1 \text{ ft}^3 \times 155 \text{ lb/ft}^3 \times (1.00 - .30 \text{ft}^3) \times .217 \text{ BTU/lb-°F}$$
$$= 23.54 \text{ BTU}$$

It can be seen that for equal storage capacity, a volume of rocks about three times the volume of water is required. The penalty for this excess volume is at least partially compensated for by simpler containerization.

An examination of Equation 3–1 reveals that for a specified heat storage capacity the designer has some choice in the material for the medium, its mass, and the temperature limits between which the system operates. The lower temperature limit is usually specified. In order that the collector efficiency remain reasonable, the upper temperature limit should not exceed this lower limit by more than about 40°F.

Losses from Storage. Losses through container walls may consume a substantial portion of the heat in storage. The amount of heat lost in this manner is given by:

$$Q_{loss} = UA\,(T_2 - T_1) \tag{3-2}$$

where

Q = heat lost through walls, BTUH

U = overall coefficient of heat transfer, BTU/hr-ft^2-°F

A = area of container walls, sq. ft.

T_2 = storage temperature, °F

T_1 = temperature of medium surrounding storage, °F

For a given storage mass, temperature difference, and U factor, heat migration through the container wall is less for a sphere than for any other shape. However, spherical containers are generally not used because of the difficulties of fabrication and insulation. Frequently, cylindrical tanks are employed for liquid systems because of the widespread availability. Otherwise, tanks having a compact rectangular plan are to be preferred. The depths of storage containers are more often than not constrained by ceiling height. For example, a container resting on the floor of a basement with an eight-foot ceiling would probably not be deeper than five feet in order to allow for top access, to limit pressure at the tank bottom, and to limit loading on the floor structure.

Heat loss from storage should be reduced by the addition of insulation. In the most benign situation, storage is located within the conditioned space. In this case, heat flow through the container is not actually a loss but an uncontrolled heat gain to the building. For storage temperatures in the range of 100°F–150°F, Anderson[3] suggests that insulation have an R value of 20. Using this reference, storage above 200°F for long periods would have up to three feet of closed cell insulation.

Location of the thermal storage system plays a prime role in the relative heat loss from storage, or conversely, in the amount of insulation necessary to maintain the heat loss at some given value. This is especially true for low temperature storage. Consider, for example, a storage system whose mean temperature is 105°F. If this system is located in a basement whose temperature is 65°F, then the heat loss will be only half as much as would occur if the system is located outside the building where the average air temperature is 25°F. (For the same environmental conditions, an inside storage tank whose mean temperature is 185°F will have three fourths the heat loss of the same tank located outside.) Conversely, twice as much insulation will be required for the outside tank in order to maintain the same heat loss. Furthermore,

Storage Materials and Construction of Container

heat lost from the outside tank is gone forever, while heat lost from the inside tank serves to decrease the building's heat demand.

Sometimes thermal storage systems are buried. The advantage of this scheme is that earth temperatures remain relatively constant year round (about the same as ground water temperature, 55°F–65°F in most parts of the United States), and that dry earth is a relatively good insulator. In order to take advantage of this constant temperature, it is necessary to bury a storage system at least eight feet below the ground surface. Although dry earth has a relatively low conductivity, such cannot be said for the same soil containing moisture. If this moisture is moving around or away from the thermal storage in any manner, increased thermal loss may be expected.

Heat Conduction in Storage. Let us define the thermal diffusivity of a material as

$$\alpha = k/\rho C_p \tag{3-3}$$

where

k = thermal conductivity

ρ = density

C_p = specific heat of material

Then, for a storage system of constant thermal conductivity the heat conduction may be written

$$\nabla^2 T + \dot{q}/k = 1/\alpha \ \partial T/\partial \tau \tag{3-4}$$

where

T = temperature

$\dot{q}$ = energy generated per unit volume

τ = time

It is understood that the Laplacian operator ∇^2 will be expressed in a coordinate system appropriate to the geometry of the thermal storage system.

Closed form solutions for (3-4) are extremely difficult to obtain. Dean[4] has used the conductive sheet analogy to approximate solutions to this and other Laplacian forms with reasonable accuracy. Although this technique saves computational effort, it is doubtful that solution of (3-4) is warranted for the design of an ordinary thermal storage system.

On the other hand, the value of α, thermal diffusivity, is of considerable interest. The larger the value of this quantity, the faster will heat diffuse through the material. On the one hand, a large value of α could be due to a large value of thermal conductivity; hence, a high rate of energy transfer. On the other hand, a large value of α could result from a small value of ρC_p the material's thermal capacity. A low thermal capacity implies that since less of the energy flowing through the medium is absorbed, then more is available for transfer. One can verify that the diffusivity of rock is about six times that of water.

Of the multitude of possible materials which might be chosen for a storage medium, water and rock have emerged as the only viable choices for the range of temperatures considered. Each has a specific heat which is acceptable. They are easily available, inexpensive, and relatively dense. Furthermore, each is fireproof, generally nontoxic, noncorrosive, and nonvolatile.

Water. Water is an almost universal choice for a thermal storage medium when liquid-cooled collectors are employed. In climates where the temperature remains above freezing, or when a drain down system for freeze protection is used, the water used for storage may be circulated through the collector loop so that it also serves as the heat transfer fluid. Otherwise, a heat exchanger between transfer and storage liquids is necessary. Care should be taken that the pressure drop through the heat exchanger be kept as low as possible so that electrical energy requirements not be excessive. When tube-in-shell exchangers are used, their mounting is external to the thermal storage. However, if open tanks or specially fabricated containers are used, heat transfer may be accomplished by copper tubing located within the storage.

Temperature variations within the storage as a result of stratification has been a subject of considerable interest. Duffie and Beckman[5] have suggested that a significant degree of stratification might occur. They model a two-section tank as indicated in Figure 3-1.

If this model is extended to an n-section tank, then the energy balance equation for the *i*th section may be written

$$(mC_p)\, dT/dZ = (\dot{m}C_p)_c\, [F_i^c\, (T_{c,o} - T_i) + (T_{i-1} - T_i) \sum_{n=1}^{i-1} F_j^c]$$

$$+ (\dot{m}C_p)_L\, [F_i^L\, (T_{L,r} - T_i) + (T_{i+1} - T_i) \sum_{j=(n-i+1)}^{n} F_j^L]$$

$$+ U_i A_i\, (T_a - T_i)$$

where the F's are step-functions defined by

$$F_i^c = \begin{cases} 1 \text{ if } T_{i-1} > T_{c,o} > T_i \\ 0 \text{ otherwise} \end{cases}$$

and

$$F_i^L = \begin{cases} 1 \text{ if } T_i > T_{L,r} > T_{i+1} \\ 0 \text{ otherwise} \end{cases}$$

The subscripts and superscripts c, L, and r refer to collector, load, and return, respectively.

When this system of equations is solved on an analog computer, the predicted system performance may differ significantly from the unstratified model. However, the use of a model consisting of more than three sections does not produce results differing markedly from three-section models.

One should note two characteristics of the model. First, the water in storage is also the liquid circulating through the collector loop. Also, the liquid circulating through the circuit satisfies demands of the load. Second, heated liquid from the collectors enters the top of storage, and the liquid returning to the collectors exits from storage bottom. Likewise, liquid to load is extracted from top of storage and its return is at storage bottom. Conditions other than these will exhibit a different stratification profile.

Close[6] has measured temperature stratification in a 41-inch high tank. Temperature measurements were taken at 6 inches, 16 inches, 26 inches, and 31 inches above

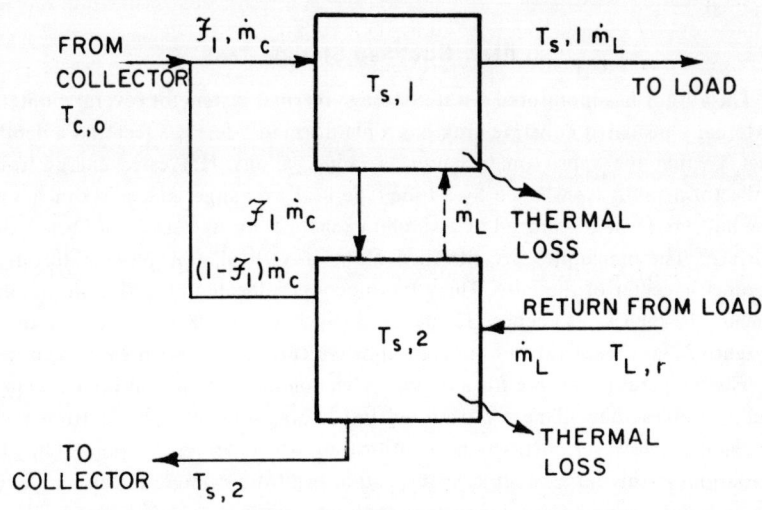

III-1. Simulated Two-Section Storage Tank

the tank bottom. At the beginning of a collection day the temperatures differed by about 25°F. At the end of a collection period this difference decreased to about 15°F, as heat was added through the collection day (Figure 3-2).

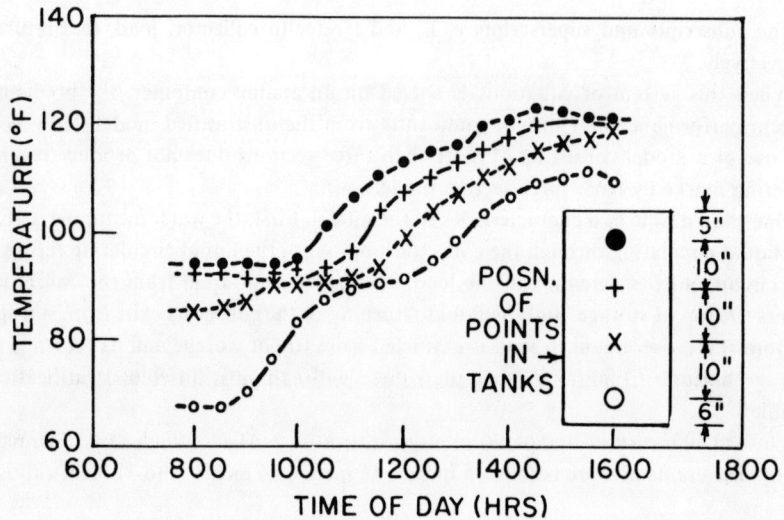

III-2. Stratified Storage Tank

The author has monitored a water storage thermal system for several months. The externally insulated concrete tank has a planform of 5 feet × 8 feet and a depth of 5 feet, so that the approximate volume is 1500 gallons. Harvested energy from the collector array is transferred by a tube type heat exchanger placed six inches above the bottom. Heat is removed by a similar exchanger located six inches below the top surface. Thermocouples are installed along a vertical axis passing through the geometric center of the base. These thermocouples located at tank bottom and at 8 inches, 16 inches, 24 inches, 32 inches, 40 inches, and 48 inches above the floor (Figure 3-3a). Temperatures are continously recorded on a strip chart recorder.

The temperature profile for a typical collection day is shown in Figure 3-3b. This curve implies that all heat introduced into storage moves upward from the heat exchanger. However, virtually no stratification occurs above this point. Only minor variations in this curve occur as heat is withdrawn from storage, and the curve retains its essential characteristics for varying tank temperatures. Taken together, these two observations suggest that any necessary supplementary heat be introduced at the top of the tank where it will remain unmixed with water in the lower portion.

Storage Materials and Construction of Container

Containers for liquid systems can be a major cost item.[7] Obviously, they must be waterproof and possess structural integrity at storage temperatures. They must be capable of resisting hydrostatic pressures which at the bottom of a five-foot deep tank will be 310 lb/ft^2. They should be capable of being drained. If an internal heat

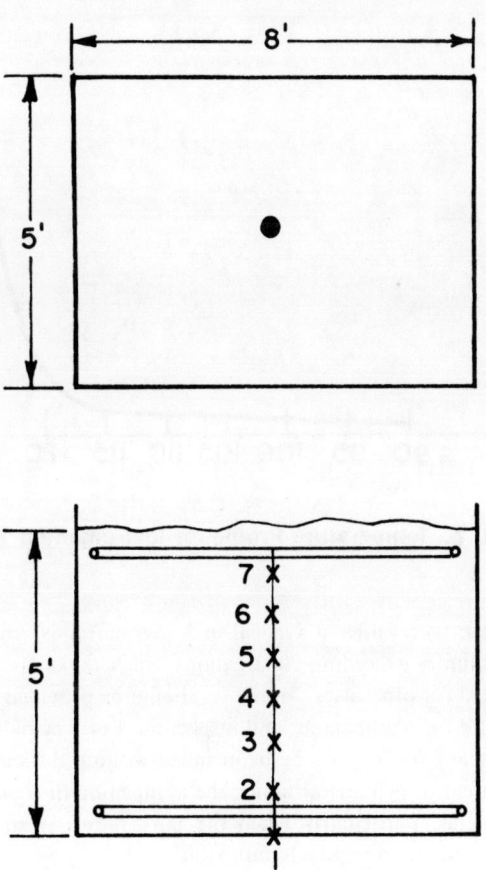

III-3. a. Instrumented Storage Tank

exchanger is employed, then some means of installing and maintaining it should be provided. They should not deteriorate, scale, or corrode in the presence of hot water, and their interior surfaces should be easily cleanable.

24 Storage Materials and Construction of Container

A spherical shape for the storage tank would be most economical of material. However, such shapes are difficult to fabricate, require more floor area than other shapes for a given volume, and may present difficulties in the installation of internal heat exchangers.

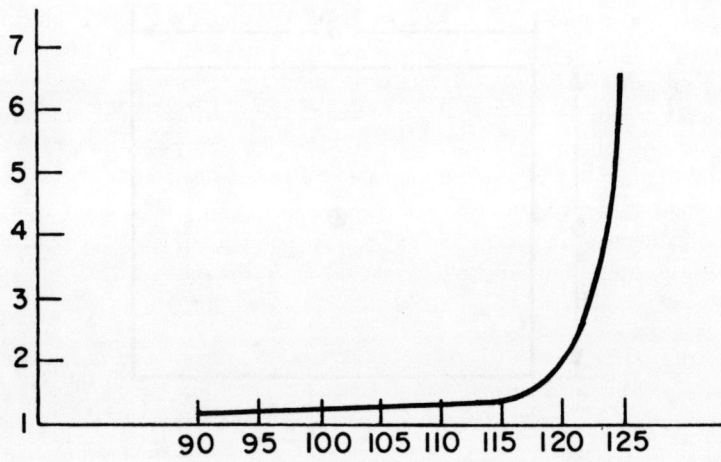

III-3. b. Temperature Profile for Instrumented Tank

Cylindrical tanks, with either a vertical or horizontal axis, are frequently used, particularly for volumes exceeding 2000 gallons. Such tanks are often available as stock items intended for other uses. Manholes should be provided for installation of heat exchangers and for maintenance and inspection. For a cylinder with a horizontal axis, internal heat exchangers may be installed without difficulty.

For either spherical or cylindrical tanks, the application of rigid board insulation will present problems, particularly when the tank radius is small. Most readily available insulation board is relatively inflexible.

Containers with vertical sides and rectangular plans will require the least floor area. When fabricated on the site they will ordinarily require the least cost of formwork. If left open on the top, access to tank interior poses no problems. On the other hand, structural requirements for the tank walls require accurate design and construction and generally necessitate the use of more material than either a spherical or cylindrical tank.

Materials suitable for tank construction include steel, concrete, and fiberglass. Steel tanks are customarily prefabricated and installed in a building during construction or buried outside. They are relatively light in weight. Piping for inlets and outlets can be installed below water level by welding, brazing, or using bulkhead fittings. Steel tanks are relatively expensive.

A few manufacturers supply fiberglass tanks. Some of these consist of two external surfaces of fiberglass separated by a core of closed-cell insulation and are therefore completely insulated before installation. However, they are expensive.

Concrete is a relatively inexpensive, easily worked material for tank construction. If the tank is to be installed during new construction, then frequently basement or footing walls may be utilized for one or more of the tank walls. Precast concrete septic tanks or cisterns having a rectangular plan and vertical walls are available in many parts of the country. Some are available with integral or internal surface waterproofing applied by the manufacturer. Piping through walls of case-in-place tanks should utilize flanged fittings of the sort used in swimming pools. Since such flanges are usually made for one and one-half inch piping, reducers may be necessary. Whenever possible, however, piping should enter through the tank top so that leakage is of no concern.

Rock. Rock is the favored thermal storage medium for air-cooled collectors. Its relatively low specific heat and the necessity for interstices for air flow require a volume about three times that of water. Nevertheless, it presents obvious advantages. It never drips through a living room ceiling. It is easy to contain. Freezing is no problem. The need for heat exchangers is removed. Consequently, collectors may operate at a higher efficiency.

On the other hand, thermal storage systems containing rock require a greater volume, and hence, a larger surface of container. As a result, heat losses from storage may be larger than for water. Depending on rock size and air flow through the bed, a disproportionate amount of energy may be consumed by the fans or blowers adequate to provide necessary air flow for sufficient heat transfer. Heating of domestic hot water from rock storage is generally restricted to a preheat function. For this application, a preheat water tank may be surrounded by the rock bed. Heat transfer from the point of contact of the rocks to the metal water storage tank is relatively inefficient. However, this heat transfer continues on a 24-hour basis.

The possibility of insect infestation and fungus growth, expecially for installations in climates with medium- or high-humidity conditions, has been inadequately investigated. Long-term use in high humidity regions may be favorable to the growth of fungus colonies which, since room air is circulated through the pebble bed, may be introduced into the lungs and respiratory systems of the building's occupants.

When thermal storage systems are used for cooling applications, existing refrigeration equipment requires temperatures of $170°F-220°F$. Rock beds do not usually store heat in this temperature range. Consequently, rock storage is useful for cooling

26   *Storage Materials and Construction of Container*

applications only in those regions which permit nocturnal cooling of the thermal storage for use as a heat sink during the following day.

Rock pile storage (also called pebble beds or packed beds) may utilize rocks of one-half to two and one-half inches in diameter. In any situation, the rock diameters should be uniform so that necessary air passages are preserved. Rocks having a small diameter present more surface to the air flow. Consequently, they are more efficient in terms of heat transfer. For example, the surface area of one cubic foot of three-inch-diameter rock is about 14 square feet, while the surface area of the same volume of one-inch rock is about 40 square feet.

Pressure drop through rock storage is a function of air velocity, rock size, and length of path. Moreover, pressure drop determines fan or blower size and energy input into the system. A rock storage having a relatively short air flow is shown in Figure 3-4a. For this system, rock diameters of one to two inches might be appropriate, since length of travel of air through the bed is only three to five feet. In comparison, rock storage requiring a long path of air travel (Figure 3-4b) might require rock sizes of two to four inches in order that pressure drop not be too great. Ideally, rocks will be small enough to effect good heat transfer but large enough to minimize pressure drop. The relationship between these factors is discussed by Löf and Hawley.[8]

The geometric shape of a packed bed storage, relative locations of inlet and outlet to collector, and relative locations of supply and return to load will determine the temperature gradient (stratification) within storage. In general, a large degree of stratification is to be desired, so that load demands may be met by air off the hottest rocks.

When either of the configurations shown in Figure 3-4 is used, air from the collector heats rocks in the upper portion of the bed to the highest temperature and these rocks

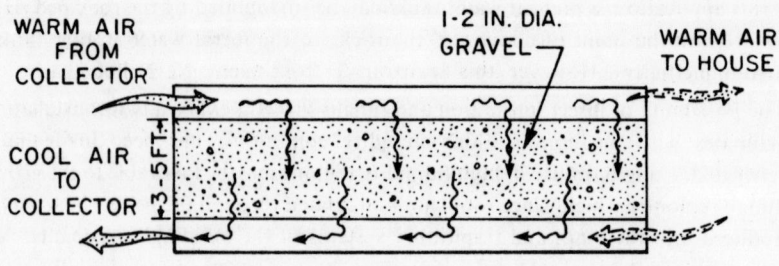

III-4. a. **Horizontal Pebble Bed**

likewise heat room supply air to the maximum. Also, air to the collector will be at as low a temperature as possible, thereby enhancing collector efficiency.

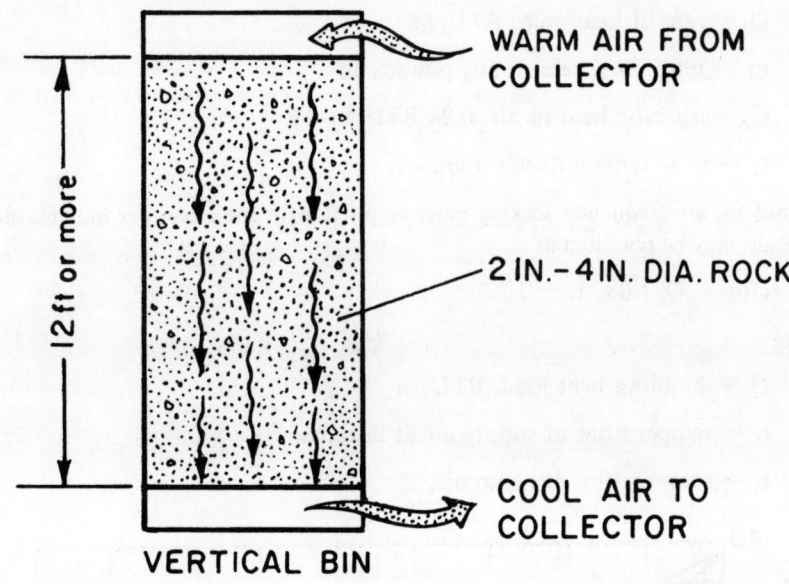

III-4. b. Vertical Pebble Bed

Some idea of the magnitude of stratification can be had from an examination of Figure 3-5. These temperature profiles were measured in one of the vertical storage units in the Löf residence. In this system warm air from the collectors enters at the bottom and supply air for the load leaves from the bottom. Nevertheless, temperature differences as high as 70°F have been recorded at different locations in the unit. Had inlets and outlets been reversed, the stratification would have been greater.

It has been stated previously that systems should be designed for a minimum storage temperature of 100°F. Many enthusiastic solar designers (and some manufacturers) prefer to rate their systems with storage temperatures as low as 75°F on the correct assumption that usable heat may still be extracted at this temperature. It is important to recognize the severe performance limitations implied by such an assumption.

Storage Materials and Construction of Container

When air is heated or cooled, its change in sensible heat is given by

$$Q = mC_p (t_2 - t_1)$$

where

Q = rate of heat gain, BTU/hr

m = mass flow rate of air, pounds/hr

C_p = specific heat of air, 0.24 BTU/lb -°F

$t_2 - t_1$ = temperature difference, °F

Since we are frequently seeking required air flow in cubic feet per minute, this equation may be rewritten as

$$cfm = Q/1.08 \ (t_s - t_a)$$

where

Q = building heat load, BTU/hr

t_s = temperature of supply air at terminal, °F

t_a = temperature of room air, °F

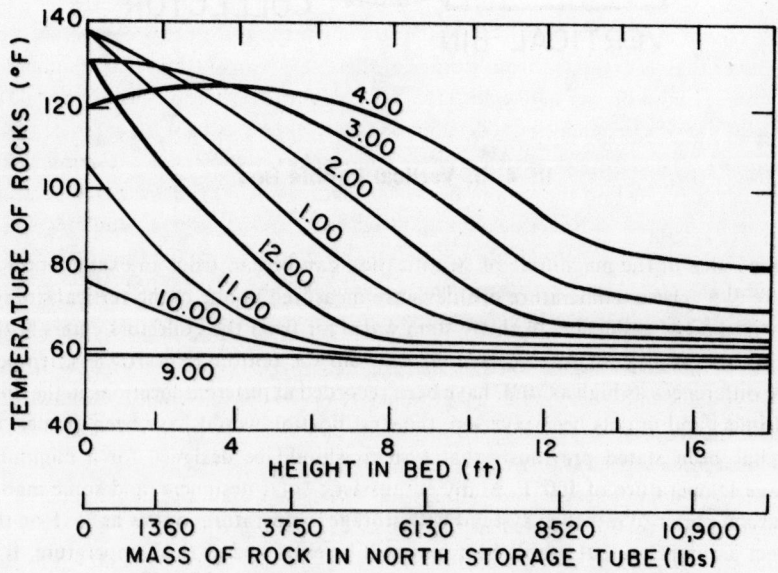

III-5. Temperature Profile for a Vertical Pebble Bed

Storage Materials and Construction of Container

Room air temperature is ordinarily taken to be 70°F. When conventional furnaces are employed, air temperature at a register is typically 140°F, so that the temperature difference is 70°F. As the register temperature decreases, so does the temperature difference, so that required air quantities are increased. Register temperatures as low as 85 °F can be tolerated provided face velocities at the register do not exceed about 350 feet per minute. Otherwise, supply air is perceived as being cold and causing drafts.

When forced air cooling is provided, a register temperature of 55°F (dry bulb) is usually specified, resulting in a temperature difference of 15°F. Although cooling loads will often be less than heating loads, a greater quantity of air is nevertheless often demanded by cooling requirements.

Let us suppose that the cooling load is 60 percent of the heating load. Then, for $t_s - t_a$ equal 15°F for cooling, a balanced $t_s - t_a$ of 25°F is necessary for heating. With this requirement, supply air must be at least 95°F and temperature of the storage at least 100°F.

It is to be expected that a low storage temperature will often occur at times of greatest heating load. Consequently, if storage temperature drops to 80°F, then air at the register may be 75°F. Either five times as much air must be delivered—an impossibility with most systems—or else heat from storage meets a maximum of only 20 percent of the building's design load. It is the latter situation which is likely to occur.

One should always be alert to the situation in which more energy is consumed by the blower than is extracted from storage at these low temperatures. For example, a 115 VAC-11 amp blower consumes 1265 W. This is the electrical equivalent of 4314 BTUh. If the blower delivers 600 cubic feet per minute against an external static pressure of 2.0 inches H_2O, then a register temperature of less than 76°F implies that more energy is put into the system than is drawn from it. This delivery temperature might be associated with thermal storage at 82°F–90°F. In this case, prudence would suggest bypassing the thermal storage system in favor of an auxiliary system even if this auxiliary uses electric resistance heat. Obviously, if natural convection is utilized, then any energy input is of no consequence.

Hybrid Systems. Although water is the usual thermal storage material for liquid-cooled systems and rock is the usual material used with air-cooled systems, it is possible to combine both air and rock in the same system or to use water as the thermal storage for air systems.

A pioneer solar experimenter, Harry E. Thomason, has ingeniously combined both water and rock in a patented thermal storage system shown in Figure 3-6.

This storage is used as part of his "Solaris" solar heating system which is also patented. Heated water from the "trickle" collector first circulates through an insulated 275-gallon tank surrounding a 42-gallon domestic water heater. From there it flows through a 1600-gallon water tank and thence back to the collectors. Since flow through the collectors is open channel, freezing is no problem. Consequently, antifreeze solutions and heat exchangers are not needed.

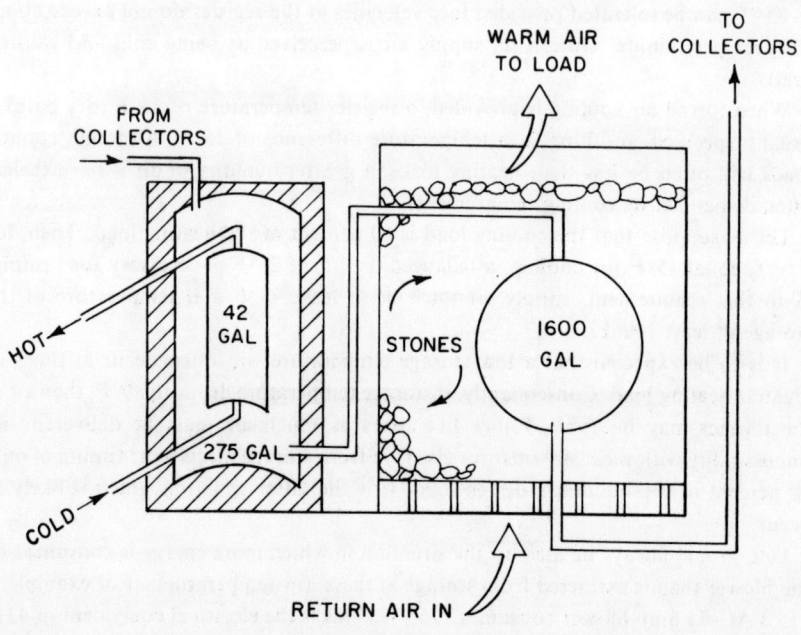

III-6. Hybrid Air and Water Storage

The 1600-gallon water storage tank is surrounded by a packed bed which is supported on hollow concrete blocks. As heat migrates from the water outward through the rocks, room return air is introduced to rock storage through the hollow concrete blocks at storage bottom and extracted at the top for room supply air. Although water is the primary storage for this system, some heat is stored in the rocks, which serve primarily as a heat exchanger.

Total Environmental Action, Inc.—a research, design and education firm headed by Bruce Anderson—has suggested using a collection of small water-filled containers as thermal storage for air collector systems. These containers, each containing one gallon or less, can be arranged in racks, on shelves, or between a building's framing members in such a way as to allow an unobstructed air flow around them (Figure 3-7).

Such a scheme offers obvious advantages. Heat storage per unit volume is about twice that of rock. A multitude of configurations is possible. When the container array is in a floor, ceiling, or interior wall location, then heat loss from storage will be a gain for the structure.

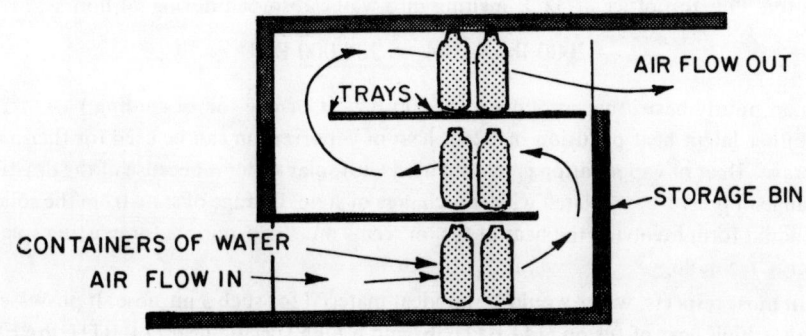

III-7. Water-Filled Containers on Racks Used with Air Systems

Performance of Thermal Storage Systems. The actual performance of a thermal storage system depends upon many factors in addition to the thermal characteristics of water and rock. The National Bureau of Standards[9] proposed a method for rating thermal storage devices. This basic method was later adopted by ASHRAE as a standard[10] in February 1977. Since ASHRAE updates its standards on a five-year cycle, it will remain in effect until February 1982.

This standard, ASHRAE 94-77, has been the subject of controversy. Some manufacturers of systems maintain that usable heat is available from storage as long as its temperature is above 70°F. It has also been pointed out that this standard is inapplicable for systems in which building components serve as some or all of the thermal storage. With a wide range of thermal storage system options, it appears unlikely that a single test procedure can be devised.

Heat of Fusion. When materials absorb (or release) energy as a result of a change in the temperature of the material, this change is the result of sensible heat storage. It is a function of the specific heat of the material for its particular phase as solid, liquid, or gas.

Additional heat is absorbed when a substance changes from a solid to a liquid or from a liquid to a gas, even though no change in temperature is measured. This heat is called latent heat. For example, 144 BTU are required to change one pound of ice at 32°F into one pound of water at 32°F. This heat is called the heat of fusion. Similarly, about 1070 BTU are required to vaporize one pound of water at 212°F into one pound of steam at the same temperature and at atmospheric pressure. This heat is called the heat of vaporization.

(The heat of fusion of water has the basis for rating air-conditioning capacity by the ton. One ton of ice at 32°F melting into water absorbed during 24 hours

$$2000 \text{ lb} \times 144 = 288,000 \text{ BTU}$$

On an hourly basis, this amounts to 12,000 BTUH or one ton of cooling.)

Either latent heat of fusion or latent heat of vaporization can be used for thermal storage. Heat of vaporization is seldom used with solar systems because of the drastic change of pressure associated with this change of state. Change of state from the solid to liquid form involving the heat of fusion seems most appropriate for existing solar energy technology.

In most respects, water would be an ideal material for such a purpose. It possesses both a high heat of fusion (144 BTU/lb) and a high specific heat (1.0 BTU/lb-°F). Unfortunately, the change from solid to liquid occurs at about 32°F, a temperature far too low for any use in ordinary space heating. On the other hand, ice was formerly used in a number of building cooling applications where building occupancy was intermittent.

The ideal heat of fusion storage material has a high heat of fusion. Its melting point will be somewhat above space design temperatures, say at least 90°F. However, its melting point will also be below that temperature at which flat plate collector efficiencies deteriorate excessively, say 140°F. There is a wide choice of materials having melting points between these two limits. However, further performance requirements will eliminate most of these materials.

Several paraffins have melting points within the desired range. Unfortunately, a sensible requirement that thermal storage not be inflammable eliminates these candidates. Other necessary performance criteria to be met include requirements that the material be nontoxic and noncorrosive, be readily available and have lowest incompatability with other materials in the system. Even when a material meets these criteria, it must meet the additional requirement of low cost imposed by economics. Cost per unit mass of technical grade material (as compared with laboratory grade) should be in the range of one to five cents per pound, when purchased in lots of a hundred pounds or more.

Materials which meet these specifications are the large volume chemicals based on compounds of sodium, potassium, calcium, magnesium, aluminum, and iron. Ideally, they will be in the form of salt-hydrates in order that advantage may be taken of the high heat of fusion of these groups. Since cost is a factor, they will probably be in the form of chlorides, sulfates, nitrates, phosphates, and carbonates.[9] Properties of five salt-hydrates are given in Table 3.

Table 3. Properties of Salt Hydrates

	Chemical Compound	Melting Point, °F	Heat of Fusion, BTU/lb	Density, lb/ft³
Calcium chloride hexahydrate	$CaCl_2 \cdot 6H_2O$	84–102	75	102
Sodium carbonate decahydrate	$Na_2CO_3 \cdot 10H_2O$	90–97	106	90
Disodium phosphate dodecahydrate	$Na_2HPO_4 \cdot 12H_2O$	97	114	95
Sodium sulfate decahydrate	$Na_2SO_4 \cdot 10H_2O$	88–90	108	91
Sodium thiosulfate pentahydrate	$Na_2S_2O_3 \cdot 5H_2O$	118–120	90	104

(From Proceedings of Workshop on Solar Energy Storage Subsystems, etc., p. 18)

Maria Telkes is likely the most knowledgeable person in the world in the field of phase change thermal storage. Her experience dates to the late 1940's when she designed a solar house for which Glauber's salt was used for thermal storage.[11,12]

Glauber's salt is the common name for sodium sulfate decahydrate ($Na_2SO_4 \cdot 10\ H_2O$) and is one of the cheapest and most easily available phase change materials. Its supply is almost inexhaustible. Although its melting point, 90°F, seems to be a little low, its relatively high heat of fusion, 108 BTU/lb, is a definite asset. The material consists of 44 percent anhydrous sodium sulfate and 56 percent water by weight, and its density is 91 lb/ft³.

At its melting point, 85 percent of the Na_2SO_4 dissolves in the water. This saturated solution has a density of approximately 83 lb/ft³. The remaining 15 percent anhydrous Na_2SO_4, which weighs 166 lb/ft³, settles to the bottom as a white sediment. For this material, it is not possible to dissolve the precipitate by raising the temperature, since the greatest solubility occurs at the melting point. If, however, the mixture is occasionally stirred, crystallization of the entire mix is obtained.

Without the presence of a nucleating agent, the temperature of the saturated liquid may drop below the freezing point without any solidification beginning. Such a situation is called *supercooling* and may be avoided by incorporating a suitable nucleating agent around which crystals may form.

The inclusion of borax, sodium tetraborate decahydrate, at a three to four percent concentration solves the problem of nucleation for Glauber's salts although the melting point is decreased to 89°F. However, the density of borax is slightly greater than that of the saturated solution, so the borax nuclei also tend to settle out.

One method of obtaining complete crystallization of the melt is to shake or invert the container in which the material is contained. This strategy, of course, is not practical for the amount of material required for any realistic amount of thermal storage. Another strategy is to encapsulate the material in containers not more than one-fourth inch deep and to rely on diffusion. For this dimension the salt crystallizes without leaving any residue. Thickening agents need to be added in order to produce a gel, from which neither the anhydrous sodium sulfate nor the nucleating agent can settle. Various organic materials, such as wood shavings, paper pulp, sawdust, and other cellulose materials, have been tried. Some of these materials work well for a number of cycles. However, they eventually hydrolize or decompose by bacterial action. This reaction may be prevented by the addition of formaldehyde or other suitable agent. Silica gel, diatomaceous earth, and other inorganic materials have also been tried. Sometimes siliceous material combines with borax, thereby inhibiting the flow of the materials into their containers. Dr. Telkes has identified a thixotropic thickening agent, attapulgite clay, which becomes quite fluid when stirred vigorously,

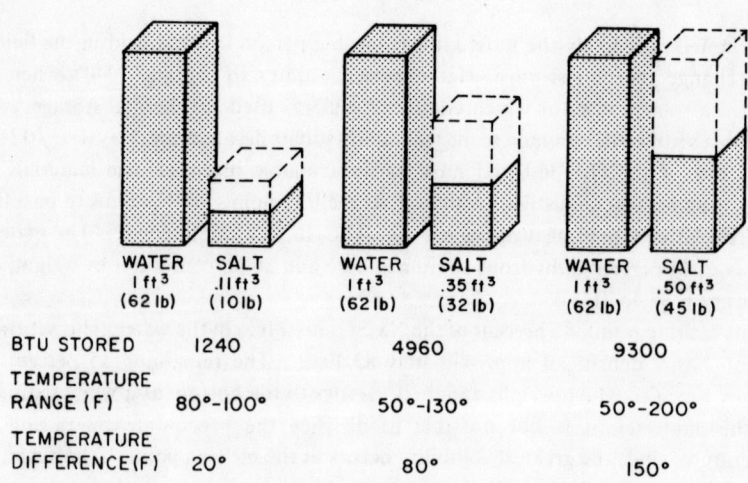

III-8. Comparison of Water and Phase-Changing Glauber's Salts

Storage Materials and Construction of Container 35

but sets to form a gel when left undisturbed. This attribute is a definite advantage when filling containers.

Containerization of materials such as these demands a minimum air and moisture migration through container walls. Metal containers have not proved satisfactory, nor are they cost-effective. Plastic containers are most satisfactory and inexpensive, but their economic viability is based on mass production. Typically, the salt-hydrates are contained in tubes whose diameters are around one and one-fourth inch or in wafers one-fourth to one-half inch thick. Several configurations are available from commercial sources. Heat of fusion thermal storage is most advantageous when the storage temperature limits are near the melting point of the material. Such a situation maximizes the effect of the heat of fusion and relies less on energy stored as sensible heat. A comparison of the effects of different temperature limits is shown in Figure 3-8.

4. PIONEERS IN THERMAL STORAGE

In most areas of this country, the average annual temperature does not differ greatly from the desired comfort temperature within our buildings. This average annual temperature, for all practical purposes, may be taken to be the same as ground water temperature, about 55°F–65°F for most locations. If one assumes neither heat loss nor gain through the building fabric, then only a modest energy input would be necessary to preserve these comfort conditions.

During much of the year ambient temperature will be at or above the comfort level for at least part of the day even though nighttime temperatures may be well below the comfort range. This situation is most pronounced in arid or semi-arid regions where diurnal variations of 30°F or more are common.

Even when ambient temperatures are low, solar radiation entering a building through transparent surfaces may be converted into heat and, if adequate thermal storage is available within the structure, can supply heat during sunless periods.

Hundreds of years ago Indians living in the desert southwest took advantage of these conditions in building their low energy habitations. Two well-known examples, Montezuma's Castle in northern Arizona and Mesa Verde in southwest Colorado, have been partially restored and are open to the public. In each case buildings are placed beneath the overhanging brow of a south-facing cliff. Low winter sun irradiates the building surfaces. During the summer months these surfaces remain in shade. Although the stone and plaster used for construction had a poor insulation value, its fairly high thermal capacity of 30 BTU/lb-°F and the huge masses used provided significant thermal storage.

As Spanish missionaries brought their culture into this region where both timber and rainfall were often in short supply, they adopted adobe construction. This material has the advantage that it is often available at the site at no cost. Its disadvantages include low resistance to moisture, poor insulation value, and the requirement of an extremely high labor input. Presumably, this last requirement posed no problem to the missionaries who, if history tells us correctly, simply impressed a few more Indians into service. Today it is often remarked that only the very rich or the very poor can afford adobe construction. Like the stone used by earlier Indians, adobe is a poor insulator with high thermal capacity. When one recalls that most adobe walls are between one and two feet thick, it is easy to see why such construction affords a significant amount of thermal storage.

The presence of such buildings and an awareness of their thermal performance has had a strong influence on the design of buildings variously known as passive, direct gain, or sun-tempered. Buildings of this sort allow radiation to enter via windows or other transparent coverings. This energy, upon striking an interior portion of the structure, is converted into heat. That portion of the heat which is not immediately used is stored in the construction. In an ideal situation (not always attainable), circulation is maintained by natural convection (thermosyphon). Such systems are distinguished by zero or minimal electric energy input and insignificant maintenance. They impose rather strong design constraints on the building form and its materials. Furthermore, until recently, insufficient data was available so that reasonable performance expectations could be met.

This type of building has a particular magnetism for those innovative architects who strive for a wholeness in design and economy of materials and systems, or (perhaps) who resent the calculations required for the design of an active system which meets specified performance criteria. It should not be taken that passive buildings cannot be designed to meet fairly rigorous performance criteria. Such criteria can be met, but the detailed engineering analysis may be more involved than that required for an active system. As it happens, many of the passive structures which have been built so far have as occupants persons who share the designer's attitudes. For these people, inside temperatures may be either below or above ordinarily specified comfort temperatures.

Solutions for the design of passive buildings are diverse. Examples which illustrate various approaches follow.

Trombe Houses. The first Trombe-Michel house was built in Odeillo, France, in 1967. Three other larger houses were built in Chauvency-le-Chateau, Meuse, in 1974. All four houses have been extensively monitored and reported in the literature.[13]

A dense cast-in-place concrete wall forms the south facade. Its outer surface is pebbled and painted either black or brown to serve as the absorber. Double glazing mounted in a metal frame is positioned about four and three-fourths inches outside the wall. Air circulation openings are positioned at the top and bottom of the wall. For the earlier house, collector area was about 16 percent of the house volume. This was reduced to 10 percent in the later houses. The concrete wall of the 1967 house is about 23⅝ inches thick; that of the 1974 houses is 14½ inches. In operation, radiation strikes the dark wall and is converted into heat. Some of this heat is used to raise the temperature of air in the interstice. The rough surface of the concrete encourages turbulence which enhances heat transfer. A natural thermosyphon is induced and mixes with room air whose design temperature is 68°F. Another portion of the heat migrates through the wall and raises the temperature of its inner surface. This heat is transferred to the room air by conduction and convection and to other interior surfaces by radiation.

Trombe reports that the Odeillo house, blessed with a sunny and mild Mediterranean climate, receives 60 to 70 percent of its heat by solar gain. The other houses, located in a less sunny region, are solar heated for 35 to 45 percent of their requirements.

Time required for heat to be conducted through the wall is critical. The thick wall requires about 14 to 16 hours, which is too long, while the 1974 walls required 9 to 10 hours, which is not long enough. Trombe suggests that for heavy masonry, such as concrete, solid brick, or rock (for which the speed of conduction is 1.5 inches per hour) an optimum wall thickness might be 15 to 17.75 inches. These suggestions are in reasonable agreement with Balcomb.[16,17]

Trombe also reports that after comparable time lapses, the inside wall temperature of the thicker 1967 wall was higher than that of the thinner 1974 walls. The reverse situation should have occurred. He attributes this anomaly to the fact that some of the heat drives off moisture in the interior of the newer wall, a situation which will certainly disappear after one or two additional years of service.

Without the use of dampers, a reverse thermocirculation can occur during night hours. Kelbaugh has solved this problem by installing flaps at the top inlets. These flaps are made of thin plastic of the sort covering garments returned from dry cleaners. Since they are top mounted, using a paper or plastic tape, and have little weight, natural convection keeps them open. However, when the heat harvest is finished, they drop due to their weight and a reverse thermocirculation is prevented.

Any of several strategies may be employed to improve the performance of this collector-storage system. Ventilating fans may be added to inlets or outlets of the concrete wall. The cover plate can be vented at top and bottom to assist in nocturnal cooling. Insulation applied to the interior wall surface could be of value. However, the most productive modification would be to add movable insulation to the outside surface of the concrete or glazing.

David Wright House. This house, located in Santa Fe, New Mexico, was designed and built by Architect David Wright for his own use in 1974, although he has since moved to California. North, west, and east walls are almost windowless and are constructed of adobe. The adobe is externally insulated with two inches of closed cell insulation and plastered. Approximately two feet of earth beneath the floor was excavated. The bottom of this excavation was insulated with two inches of closed cell insulation, backfilled, and finished with a brick floor. Additional thermal mass was incorporated in water-filled 55-gallon drums which serve as fill for adobe covered bancos, built-in benches. Liquid-cooled collectors provide domestic hot water and a Franklin stove provides auxiliary heat. The house is an open plan with gently curving interior walls which allow for minimum obstruction to natural convection.

The entire south wall of the building is glazed with two panes of tempered glass. Vertically operating shutters of two-inch insulation are provided to prevent heat loss during nocturnal periods.

This house is one of a series designed by Wright, some of which incorporate an inside swimming pool for thermal storage.

Atascadero House. Harold Hay conceived the idea of collecting, storing and transferring solar heat by use of roof-mounted water storage containers. In 1967 he constructed a solar test room with John Yelliot and subsequently patented the concept. A full-sized house utilizing the idea was constructed in Atascadero, California in 1972 (Fig. 4-1).

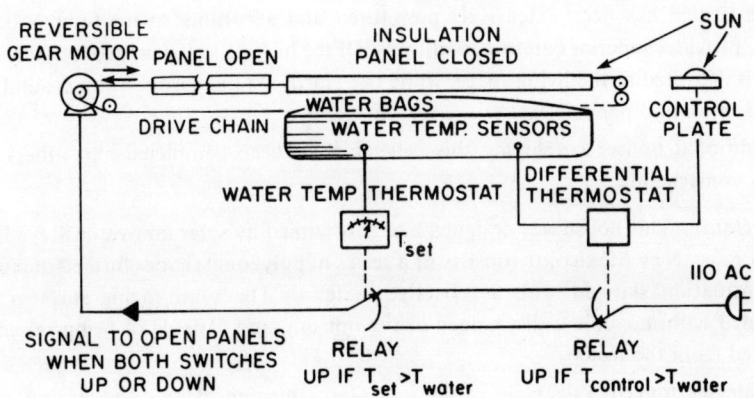

IV-1. Atascadero House, Roof Section

This house contains 1100 square feet. Exterior walls are 40 percent frame construction with three-inch fiberglass insulation; 25 percent double-glazed windows and doors; and a 35 percent combination of 12-inch thick concrete block with no insulation and eight-inch thick concrete block with vermiculite-filled cavities. Interior load-bearing walls are eight-inch concrete block with sand-filled cavities. Perimeter insulation is two inches thick.

The entire metal ceiling of this structure is covered with clear polyvinylchloride (PVC) bags containing an eight-inch depth of water. An inflatable air cell is incorporated above the water. These bags may be covered by sliding over them two-inch thick polyurethane insulation panels which may be otherwise stored in a three-level stack above the carport.

To collect heat, the bags are exposed by storing insulating panels above the carport. When radiation is unavailable or unwanted, these panels cover the bags. For summer cooling the bags are exposed at night to permit radiative and convective cooling to the night sky.

Water in the bags serves as a heat source or sink. In the heating mode, heat in the water is conducted to the steel ceiling where it is introduced into the interior spaces by convection and radiation. An opposite situation occurs during the cooling mode. No humidity control is provided.

Movement of the insulating cover panels is effected by a reversible gear motor actuated by differential thermostats. Although such a mechanism effectively identifies this scheme as an active system, the energy input is modest. Furthermore, moving the insulating shutters could be a manual operation.

The system has been extensively monitored and according to the house's occupants, provides superior comfort conditions. Of the heat collected, approximately one third is delivered to the living space, while two thirds are lost through and around the closed insulation panels.

Additional houses employing this scheme have been completed and others are under construction.

Baer House. This house was designed and is occupied by solar innovator Steve Baer in Corrales, New Mexico. It consists of a series of polygonal shapes formed of closed cell insulation skinned with a reflective material. The south-facing surfaces are provided with movable walls, hinged at the bottom, and capable of being raised or lowered from the inside.

Inside the hinged walls is an expanse of plastic glazing. Behind the glazing is an array of 55-gallon drums compactly supported by steel framing. These sealed drums are filled with an antifreeze solution and are painted black on their irradiated surfaces.

During periods of heat collection the wall is lowered. Its reflective inner surface reflects additional radiation to the glazing. The drums, acting as absorbers, transfer heat to their contained water. In turn, this water heats interior spaces by the usual methods of heat transfer.

Baer has applied this principle to other buildings. Bill Yanda, working with Baer, has applied the Drumwall concept as thermal storage for passive greenhouses.

Dean House. Outwardly, the Dean house typifies the active approach to solar space heating.[15] It is, in fact, a hybrid with the passive elements incorporated first. Only then were sufficient liquid-cooled collectors, storage, and a heat pump system (described in Chapter 5) installed to guarantee performance exceeding that normally encountered in conventional systems (Figure 4-2).

The dwelling contains approximately 1900 square feet, a 750 square foot basement, 240 square foot greenhouse, and garage, storage, and porch areas. Outside design temperature is $-5°F$ and inside design temperatures for residence, basement, and greenhouse are $70°F$, $65°F$, and $50°F$, respectively.

The house is zoned with coolest areas on the north. All entrances are equipped with air locks. Exterior walls are two inches by six inches with five and one-half inches of fiberglass insulation. All windows are triple-glazed and fitted with interior insulating shutters. North and west walls are sunk four feet into the earth. These walls, as well as basement and greenhouse walls, are externally insulated with waterproofed two-inch closed-cell insulation. The five and one-half-inch concrete floor and basement ceiling is surfaced with two and one-fourth-inch hard-burned dense paving bricks. A 16-inch by 40-inch masonry flue serving the wood-burning stove is exposed in the basement as well as throughout both floors of the house. Altogether, the 60 tons of insulated thermal mass is capable of storing about 24,000 BTU/°F temperature difference.

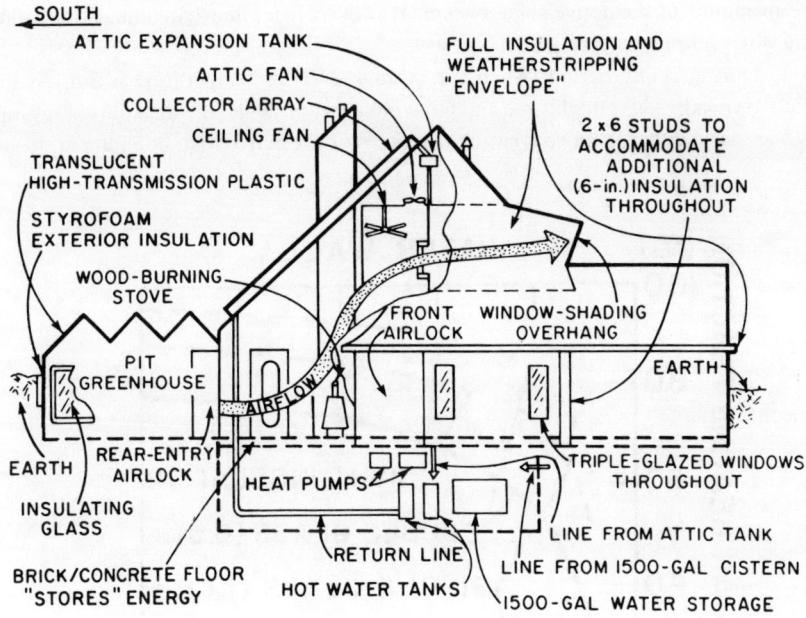

IV-2. Section of Dean House

The attached 12 foot by 20 foot pit greenhouse has its north wall in common with the residence. Double-glazing is employed throughout, with a suitable fiberglass reinforced plastic used for south-inclined roof surfaces and glass on wall surfaces. Raised earth planting beds two feet wide and four feet high are used for additional thermal storage.

A reversing thermostatically controlled fan is installed between greenhouse and dwelling. This fan consumes 68 watts. When greenhouse temperature exceeds 80°F, then this fan supplies heated air to the dwelling. When greenhouse temperature is below 80°F but above 70°F, then an equilibrium condition is assumed to exist and the fan is inactive. When greenhouse temperature drops below 70°F, then air from the house is used to heat the greenhouse. During the extreme winter of 1977 when ambient temperatures plunged to −18°F the greenhouse temperature remained above 50°F. This allowed many vegetables to be harvested throughout the entire winter.

The building is modestly instrumented. Five separate electric meters have been installed by the local utility. Ten thermometers are strategically located. Twelve thermocouples transmit data to a recorder. Most of these devices are used to monitor performance of the active solar system. However, from hourly readings, the following observations may be made:

1. During sunny periods when the average 24-hour temperature is not less than 45°F (typically, 30°F nighttime and 60°F daytime, or 20 degree-days) no energy input is necessary, either from conventional sources or the active solar component. In such

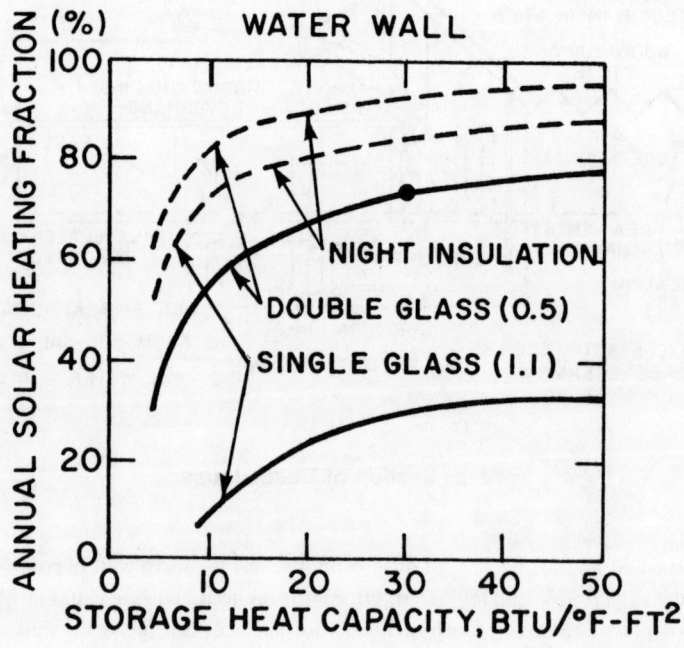

IV-3. a. Effect of Storage Mass

cases, inside ambient temperatures remain between 71°F and 76°F, while greenhouse temperatures vary from 60°F to 85°F.

2. Regardless of outside temperature, if the house is sealed, then it will exhibit a temperature swing of less than 5°F during any 24-hour period with zero energy input. This observation has obvious implications as on- and off-peak utility pricing becomes more widespread.

Performance Verification for Passive Buildings. Each of the buildings described above represents a different approach to passive design. It would be unfair to suggest that one is better than another. Yet it is likely that these buildings, with the possibility of the Trombe designs, were not a priori blessed with any sort of performance analysis. Each designer used a combination of his technical knowledge, experience, and subjective bias to create what he thought best. For those designers who live in their own buildings, or for those other occupants whose ecological or philosophical

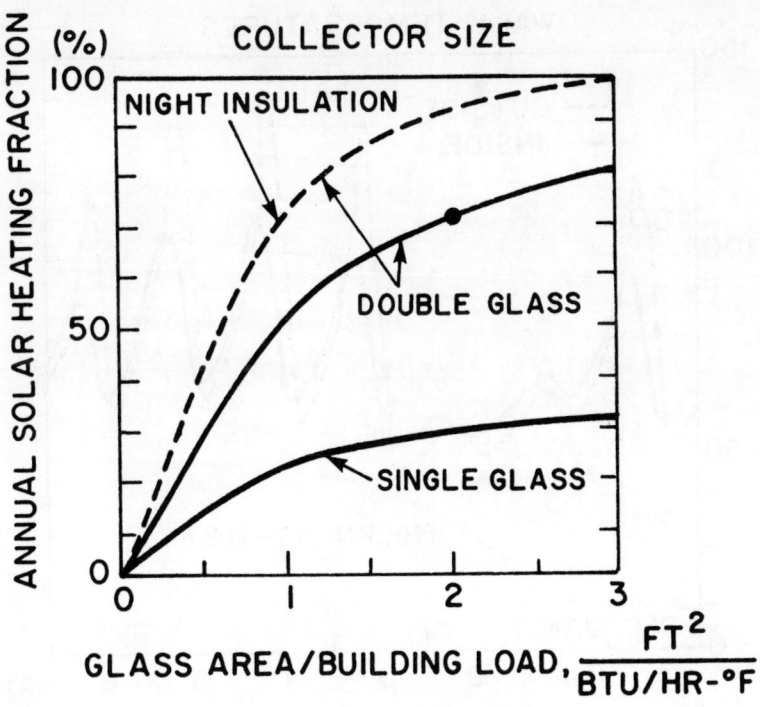

IV-3. b. Effect of Glass Area

positions are in accord with the designers of these buildings, no particular problem is presented. But, one wonders, can the performance of passive architecture be predicted well enough to satisfy the needs and desires of mainstream America?

This question doubtless occurred to Balcomb and his co-investigators at Los Alamos Scientific Laboratories.[16] Under the auspices of ERDA they monitored eleven passive houses, ten of which are in New Mexico. The eleventh is in New Jersey. They constructed and monitored two test rooms.[17] Finally, they employed the considerable scientific talent and computational capabilities of LASL to evaluate the performance of passive solar-heated buildings.

Their investigations concerned thermal mass located a few inches interior to south-facing glazing. Thermal mass consisted of heavy masonry or water contained in vertical axis fiberglass-reinforced plastic cylinders having a 12-inch diameter. In the test rooms, masonry was 16 inches thick and water cylinders had a 12-inch diameter.

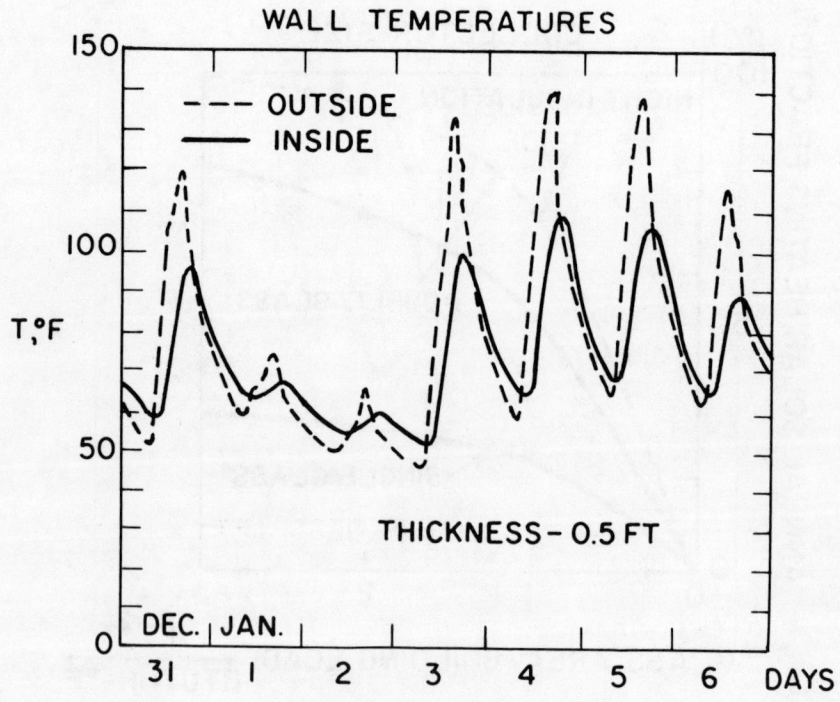

IV-4. a. Time Response for Six-Inch Masonry Wall for
One Week Period

Data obtained from these test rooms were correlated with meteorological data for 29 selected points around the world and for particular years. Simulations were calculated for the 8760 hours of each year. Subsequently, the annual contribution due to solar was calculated.

A summary of their conclusions—adequate for conventional passive solar design—includes:

1. With a water wall, daily interior temperature variation on a clear day is approximately 20°F. For an unventilated masonry wall this variation is 11°F and for a wall with vents open might be 24°F.

2. During sunny periods, average room temperatures were 60°F to 70°F above average ambient. On a winter day when the average ambient temperature was 23°F, the room with a water wall had temperatures of 74°F to 94°F. The room with a

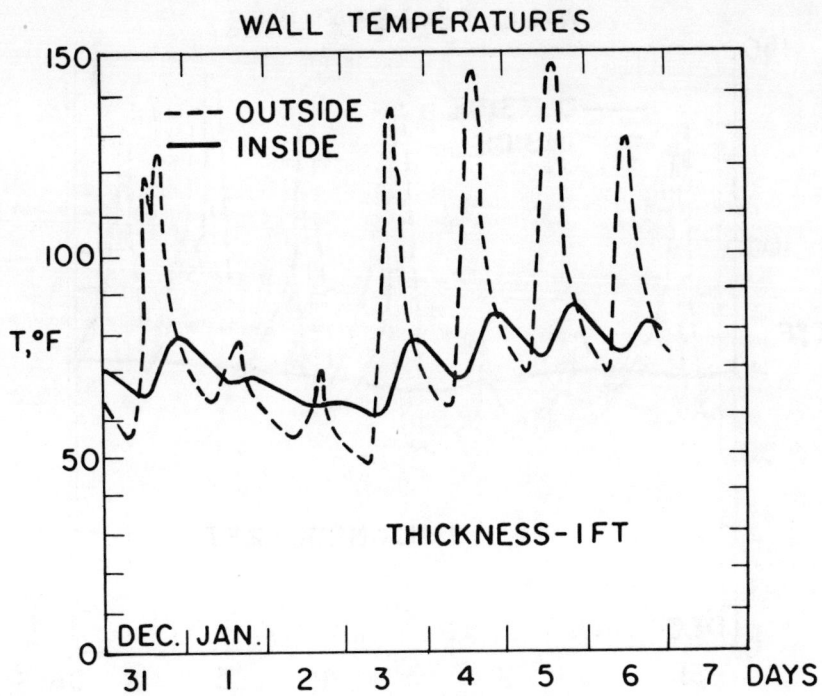

IV-4. b. Time Response for Twelve-Inch Masonry Wall for
One Week Period

masonry wall had interior temperatures of 84°F to 96°F. The top vent temperature of the masonry wall reached 132°F. During a cloudy period in which the ambient temperature held at 20°F, temperatures of both rooms dropped to 48°F (Figure 4-3).

3. Single-glazing is a poor choice. Double-glazing results in much better performance, but night insulation makes single-glazing a viable option and double-glazing even better.

4. Given the same thermal capacity and glass area, a water wall will slightly outperform a masonry wall.

5. A thermal storage capacity of at least 30 BTU per square foot of glass is necessary to reduce undesirable temperature variations within the heated space.

6. For a solid wall, thickness has considerable significance. If the wall is six inches

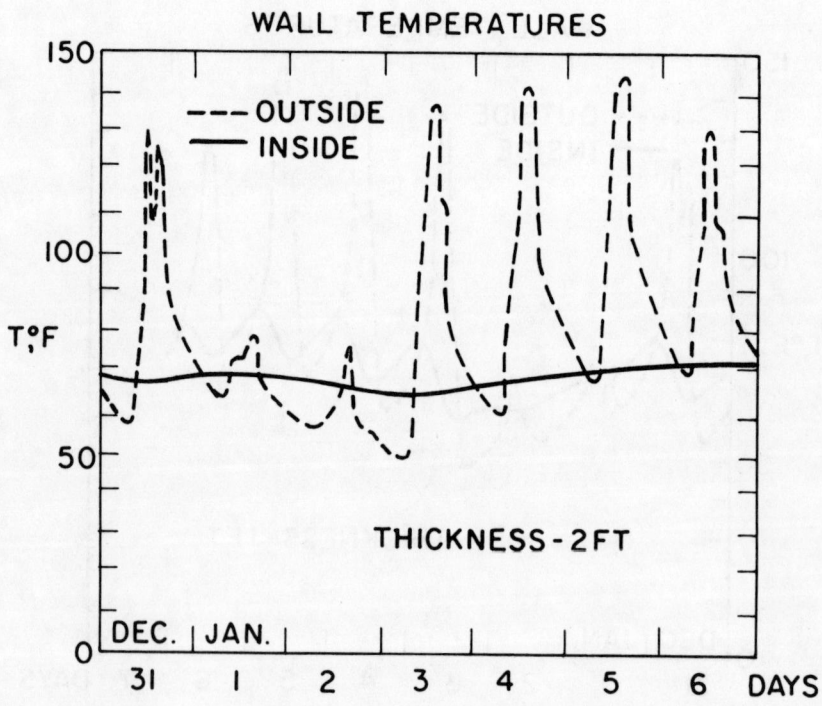

IV-4. c. Time Response for Twenty-four Inch Masonry Wall for One Week Period

thick then interior temperatures may fluctuate widely. With a two-foot-thick wall inside temperature changes are more moderate, but response time is much longer. The reference suggests an optimum thickness of 12 inches, somewhat less than that recommended by Trombe.

7. Vents for thermocirculation such as those used by Trombe will increase the overall annual performance provided that dampers to prevent reverse flow are incorporated.

There is little question that passively heated buildings offer an exciting alternate to more conventional design. However, the designer of such buildings should be ever watchful lest storage walls interfere with the placement of ventilating windows which may be used for summer cooling and be aware that external shading may be necessary to prevent walls from overheating in summer.

5. THERMAL STORAGE STRATEGIES

It has been mentioned earlier that water and rock emerge as the two best materials for the storage of sensible heat when all factors are considered. Even with this limited choice of materials, a substantial array of applications is available. Their use in passive systems were discussed in the last chapter.

As long as 75 years ago, notably high temperatures were observed at the bottoms of a group of Hungarian lakes. This phenomenon served as the basis for the concept of the "solar pond," one of whose first investigators was Harry Tabor[18,19] of the National Physical Institute of Israel. The physics of such ponds has been described by Weinberger.[20] Current research projects are underway in this country as well as elsewhere.

A solar pond consists of a clear body of water that is sufficiently shallow to allow a large portion of solar energy to penetrate to the bottom (often painted black) where it is absorbed. In a homogeneous pond the heated water expands, rises to the top, cools, and sinks again. This results in a pond possessing limited stratification. However, the solar pond consists of two layers of water. The top layer is fresh water while the lower layer contains a high salt concentration. The thermal expansion of the water heated at the bottom is insufficient to disturb the stability provided by the salt concentration so that convection is minimized. Since water is opaque to long-wave (heat) radiation, the only heat lost to the atmosphere is by conduction. However, water is a relatively good insulator so that this loss is minimal. A schematic of a solar pond is shown in Figure 5-1.

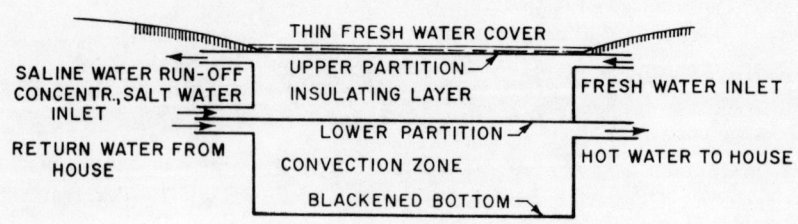

V-1. A Solar Pond Collector

Although such ponds may require as long as a year to reach a steady temperature condition, this temperature may be as high as 190°F.[21,22]

Problems associated with solar ponds include: (1) mixing of the salt and fresh water, (2) evaporation, (3) reflections from waves, (4) biological growth, and (5) decrease of transparency due to dirt.

These problems should in no way be considered insurmountable. Mixing of fresh and salt water may be nullified by periodically replenishing the fresh water in the top layer and adding salt to the lower. Mixing may be avoided completely by separating the two layers of a thin transparent plastic film. Evaporation may also be controlled by this method. Inhibition of biological growth may be effected by chemical treatment. Wave action may be minimized by judicious location of the pond and wind breaks.

While not as efficient as a solar pond, a swimming pool can be used for thermal storage. Surface losses due to wind and evaporation tend to be excessively high for outdoor pools unless a cover is employed. David Wright incorporated an indoor pool in the passive house for Karen Terry. This pool serves as thermal storage. However, if swimming pools are also to be used for their usual purpose, it is unlikely that the water temperature is high enough to be used for space heating in either hydronic or forced-air systems. Nevertheless, swimming pools as well as solar ponds can be used with heat pumps. Furthermore, when a heat pump is properly integrated into a "conventional" active solar energy system, the over-all operating cost is less than either the heat pump or solar (with auxiliary) acting alone. In addition, the usable capacity of a given storage system may be doubled, or even tripled.

The heat pump was long considered an engineering novelty. Although conceived by Lord Kelvin in 1852, its first practical installation was made in Scotland only in 1927. The heat pump enjoyed a brief popularity in the U.S. during the 1950's but was plagued by reliability and control problems. Both reliability and control have been remarkably improved. Today most manufacturers of refrigeration equipment supply a range of heat pumps for the majority of applications. The efficacy of heat pumps in the future is generally recognized by informed persons in the United States. Except in the coldest weather they are far more efficient than electric resistance heating. The rising cost and possible unavailability of natural gas and fuel oil enhance the cost effectiveness of heat pumps. Although coal reserves are adequate for many years, the delivery of coal to most homes in the United States is a formidable task. At this time legislation is in process to encourage users to install heat pumps.

A heat pump operates on the thermodynamic principles of the mechanical refrigeration cycle which is related to the ideal, or Carnot, cycle. Its ideal coefficient of performance (C.O.P.) is given by

$$\text{C.O.P.} = T_1/(T_1 - T_2)$$

where T_1 is the condensing temperature and T_2 the evaporating temperature, both expressed in degrees Rankine. From this expression it is obvious that the most economical operation of a heat pump occurs when the difference in temperature between condenser and evaporator is as small as possible. This fact is central to solar assist.

A commercial heat pump used for cooling operates as any other refrigerating machine. Heat is absorbed from room air into the cold evaporator and rejected to outside air or water from the hot condenser. When the heat pump is used for the heating cycle, heat is absorbed from the outside air or water by an even colder evaporator. This heat, along with the heat of compression, is rejected to the air of the conditioned space.

In regions of mild winter climate where a small temperature difference exists between evaporator and condenser, the heat output theoretically may be more than six times the heat equivalent of compressor work. In actual practice the Carnot cycle is not attained, but a heat output of two to four times the compressor work may actually be realized in mild winter climates. For a real heat pump, its actual coefficient of performance may be defined as:

$$\text{C.O.P.} = \frac{\text{heat obtained from the condenser (heating coil)}}{\text{heat equivalent of electric energy input to compressor motor}}$$

Table 4. Performance of a Commercially Available 3-Ton Heat Pump

Outside Temperature °F	Capacity 1000 BTUh	Current Draw kW	Coefficient Of Performance
70	46.0	3.8	3.55
60	39.0	3.5	3.27
50	33.0	3.3	2.93
47	31.0	3.3	2.75
40	27.5	3.1	2.60
30	22.0	2.8	2.30
20	18.0	2.6	2.03
17	17.0	2.6	1.92
10	14.5	2.5	1.70
0	11.0	2.3	1.40
−10	7.5	2.0	1.10

Table 4 indicates the actual capacity and C.O.P. of a small high-quality heat pump manufactured by the Carrier Corporation. An examination of this table reveals two unique aspects of heat pump systems. First, the heating capacity of a system decreases with decreasing outside temperature. Low outside temperatures generally imply the need for a greater heating capacity. Consequently, when such systems are installed in regions where extremely cold weather persists they are usually fitted with supplemental electric resistance heaters.

It has been noted that the C.O.P. decreases with decreasing outside temperature. With an outside temperature of 50°F, about 3 BTU are supplied to the space for each BTU supplied by the electric utility grid. On the other hand, with an outside temperature of −10°F, the system performs hardly better than electric resistance heating used alone. Taken together, these two reasons account in part for the favorable acceptance of heat pump systems in locations experiencing mild winters. Heat pump systems are frequently avoided elsewhere.

Water-to-air heat pumps are not subject to such wide variations in capacity and C.O.P., since they are independent of outside air temperature. However, since the water-to-air heat pump requires a source of water in a specified temperature range, it is seldom used in small installations. Its usual application has been in large multizone buildings where simultaneous heating and cooling might be called for and the cost of boilers and chillers or cooling tower can be justified.

In order to understand the synergistic heat pump-solar relationship, the following points are summarized:

1. For space heating using either a forced air or hydronic system, storage temperature should not be less than 100°F.

2. The efficiency of a flat plate collector declines as the temperature difference between ambient and absorber increases. During extremely cold weather this difference may be so great that the contribution from a single glazed collector may be almost worthless. Double or triple glazing will help this situation, but with an increase in initial cost.

3. The C.O.P. (efficiency) and capacity of a heat pump decrease as the fluid (air or water) temperature across the evaporator decreases. For the typical air-to-air heat pump used in residential and small commercial applications, this situation requires the use of supplementary electric resistance heaters.

An effective solar-heat pump system is illustrated in Figure 5-2. This system employs a water-air heat pump of the sort produced by several manufacturers. Typically, these units require water between 60°F to 95°F year-round. For this range, C.O.P. remains essentially three and capacity is fairly constant. For these units, E.E.R. ranges from 7.5 to 9.5.

With single glazing, it is easy to maintain storage temperature above 100°F during the early part of a heating season.

During this period, solar heated water is circulated through a finned tubular hot-water coil in the return air plenum. Only the heat pump blower is used. During the middle of the heating season, mid-December to mid-February, ambient temperatures are low enough that to maintain tank temperature above 100°F would result in a severe impairment of collector performance. Consequently, during this period, storage temperature is deliberately allowed to drift downward to the 60°F to 90°F range and collectors are permitted to operate at an enhanced efficiency. Controls for a system such as this may be as automatic as desired, using existing "off the shelf" items common to hydronic installations.

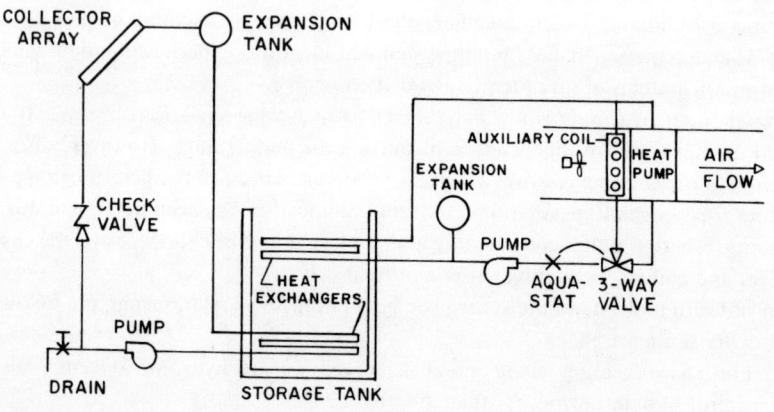

V-2. Solar Assist for a Water-to-air Heat Pump

In much of the interior United States characterized by continental climatic patterns, sudden and violent weather changes are frequent. Deep-winter temperatures vary over a wide range and periods of sunny and cloudy weather are erratic and unpredictable. Consequently, maximum storage temperatures of 140°F may be possible in January. If thermal storage consists of 1000 gallons, then, with a "conventional" solar system, 333,200 BTU would be available for heating and it would be necessary to purchase any additional needed energy.

Suppose this same storage system served as solar-assisted heat pump with a C.O.P. of three. The same amount of energy would be extracted in the same manner until the storage temperature dropped to 100°F. At this point, the storage assists the heat pump, providing two BTU for every one BTU of purchased energy until, at 60°F, the storage is exhausted. At this point 833,000 BTU will have been delivered to the space.

The same example can be viewed from another position. Suppose 833,000 BTU need to be delivered with a starting storage temperature of 140°F. Assume that electric resistance heat is used for backup and that this costs $.06 per kWh. For the straight solar installation, 499,800 BTU equivalents must be purchased at a cost of

$8.79. For the heat pump system, 166,200 BTU equivalent will cost $2.92, one third as much. When it is realized that such a situation may occur several times during a season, the efficiency of the solar heat pump team may be well appreciated.

When summer cooling is desired, this sort of system requires some kind of heat sink. This sink is usually a cooling tower or pond, a deep well, or a heat exchanger in a lake or stream. It should be possible to devise a north-facing unglazed collector for nocturnal cooling. The necessity for these rather elaborate heat sinks is the reason that water-to-air heat pumps are infrequently used for small applications.

On the other hand, literally hundreds of thousands of air-to-air heat pumps are presently in use throughout the country. Some of the structures which they serve are prime candidates for a solar assist retrofit. A scheme embodying the same principles described above is shown in Figure 5-3.

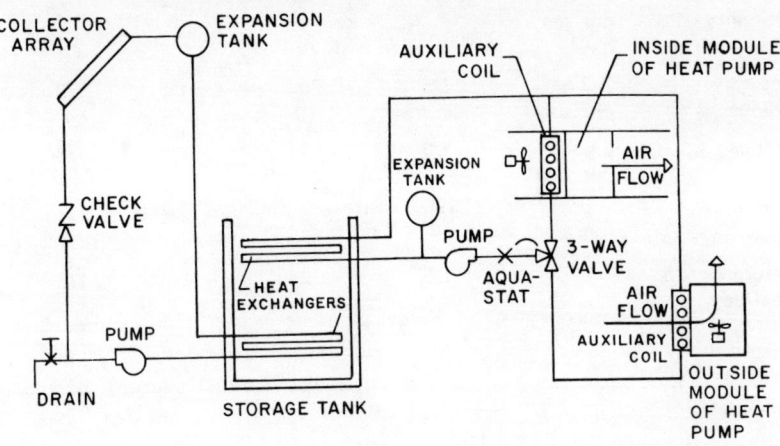

V-3. Solar Assist for an Air-to-air Heat Pump

There is no reason why pebble beds cannot also be used to assist heat pumps in the manner just described. One should remember that the propellor fan used with outside units is capable of moving large quantities of air, but will not operate satisfactorily against much external static pressure. Since the static pressure drop through a pebble bed is significant, an auxiliary blower will probably be needed.

Thermal Storage for Domestic Hot Water. Solar assist for domestic hot water has the most rapid amortization of any flat-plate solar system, since the solar supplement may be used year round. Furthermore, in this case, the storage tank is the reservoir from which hot water is drawn.

In order to design such a system, some estimate of hot water demand is necessary. Table 5 indicates one method for sizing a hot water system when heat is available from conventional sources.

**Table 5. Hot-water Demand Characteristics.
Adapted from *ASHRAE Guide and Data Book 1962.***

Type of Building	Hot Water Required Per Person	Max. Hourly Demand in Relation to Day's Use	Duration of Peak Load Hours	Storage Capacity in Relation to Day's Use	Heating Capacity in Relation to Day's Use
Residences, apartments, hotels	20 to 40 gal. per day*	1/7	4	1/5	1/7
Office buildings	2-3 gal. per day*	1/5	2	1/5	1/6
Factories	5 gal. per day*	1/3	1	2/5	1/8
Restaurants 3 meals/day		1/10	8	1/5	1/10
Restaurants 1 meal/day		1/5	2	2/5	1/6

*At 140°F.

It can be visualized that a large storage capacity can result in smaller heating capacity. It is also possible to manipulate calculations when using this table so as to reduce heating capacity while simultaneously reducing storage size. This fact suggests that data in the table are on the conservative side.

Actual hot water use in a residence is subject to wide variation depending on personal habits, how and if a dishwasher is used, and whether clothing is washed in hot, lukewarm, or cold water.

In the usual (nonsolar) installation, the water heater is chosen for its storage capacity (30 gallons to 60 gallons, or more) and its recovery rate. The recovery rate is the quantity of water whose temperature can be raised 100°F in one hour. A 3800-watt dual element electric heater may have a recovery rate of 15 gph. Temperature in the tank is usually maintained at 140°F to 150°F. Residential dishwashers are designed to use 140°F water and have an internal booster heater to elevate water to a sanitizing temperature. Otherwise, hot water used in the home is usually from 95°F to 105°F according to personal preference. The lower temperature is obtained by mixing at the tap.

If one has a private water supply—a deep well, for example—then the entering water temperature may have a relatively constant temperature corresponding to that of the local ground water. If, on the other hand, water is drawn from a distribution system, then its temperature should be determined according to measurement. In this case, the annual variation will be determined by depth of the water main, method of storage (elevated or buried), proximity of storage to users, and rate of use by customers.

Some "rules of thumb" have evolved as a result of these variables. For a family of four persons, the solar preheat tank should not have less than an 80-gallon capacity, with another 20 gallons for each additional family member. In general, 17 to 20 square feet of collector—about one commercially available panel—should be allocated for each 20 gallons of storage. The most simple application is shown in Figure 5–4.

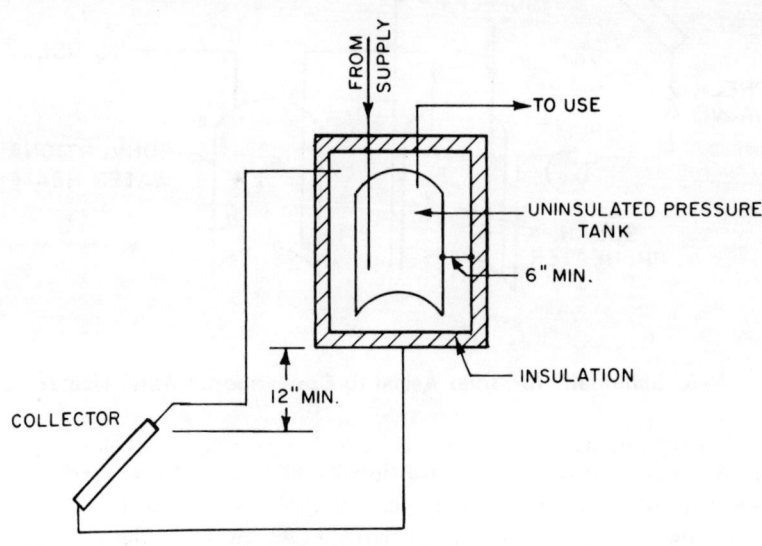

V-4. Thermosyphonic Domestic Water Heating Schematic

This system relies on thermosyphonic circulation for its operation. An uninsulated preheat pressure tank is surrounded by a water jacket containing liquid from the collector loop. About six inches of liquid should surround the pressure tank and the bottom of this liquid should be at least 12 inches above the top of the collectors for the system to operate. Obviously, the outside of this liquid should be insulated in order to maintain as high a temperature as possible. Thermosyphonic circulation is decreased as circuit piping diameter decreases, as total length increases, and as total elevation difference decreases.

The advantages of a system such as this are that no pumps or controls are involved nor, if an antifreeze solution is employed, is any routine attention required. On the other hand, additional structure may be required to support the storage tank. Furthermore, heat transfer between the collector loop and storage may be difficult to predict.

The circuit diagram for solar preheat as it is described in most reference texts is shown in Figure 5-5.

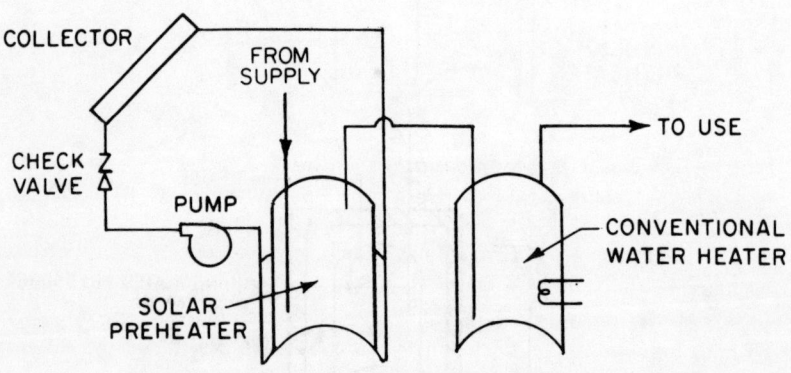

V-5. Schematic for Solar Assist to Conventional Water Heater

In this diagram, water from source (usually 40°F to 70°F) is accepted by the preheat tank where it is heated as much as possible by the solar assist. From this point it flows through a conventional water heater in which the temperature is elevated to a service value.

Although such a system has merit, it suffers from a major drawback. During much of the year, a collector array is capable of independently raising preheat tank temperature to the service level. From this point, hot water flows into the conventional heater where it will cool if not used immediately, causing the tank heater to energize. Although not generally recognized, the average 40-gallon domestic water heater requires about 10^6 BTU/month (293 kWh/mo) to maintain water at service temperature. This may partially account for the difference between some electric utilities' estimates of usage for this purpose and their customers' actual bill. One method of reducing this difference is to install a pair of gate valves in the system as shown in Figure 5-6.

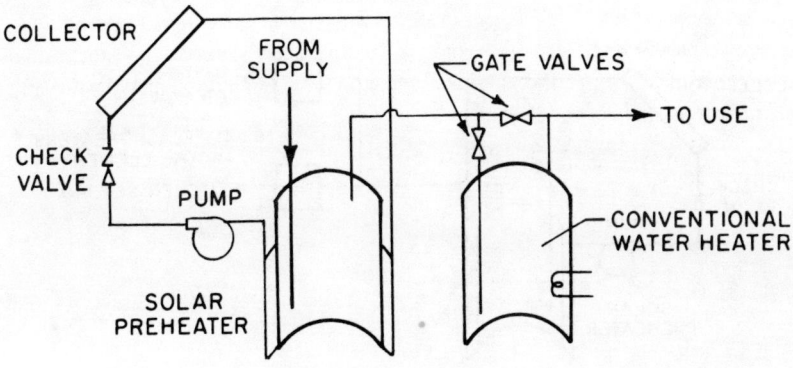

V-6. Modification of Figure V-5 for Minimizing Conventional Energy Input

With this arrangement, the conventional heater can be completely "short-circuited." During much of the year after sunny days, the conventional heater is completely unnecessary if the dishwasher is used during afternoon or early evening hours and bathing follows. Adequate warm water will still be available the following morning for normal use.

The author's home, using an 82-gallon preheat tank, uses this system. During the five-month nonheating season, the system has operated with zero energy from conventional sources and with an abundance of hot water, if used during the appropriate times (i.e., late afternoon and evening). The disadvantage of this system is the need for occasional manual or automated return to a preheat mode.

A more energy efficient system meeting the performance and maintenance specifications of the general public has been devised by the author. Shown in Figure 5-7, it completely eliminates the conventional 40- or 50-gallon water heater with its attendant storage losses.

In this system, service hot water is heated as much as possible in the solar preheat tank. From this point it flows to the kitchen, bath, or other location. At each point of use is located a small, six-gallon "flash" heater with 4500W heater. The thermostat in the kitchen location is set at 140°F. In other locations, it is set at 120°F or lower, or turned off completely. In this manner, standby losses are minimized. From the viewpoint of initial cost, two or three flash heaters cost about the same as one conventional storage tank heater.

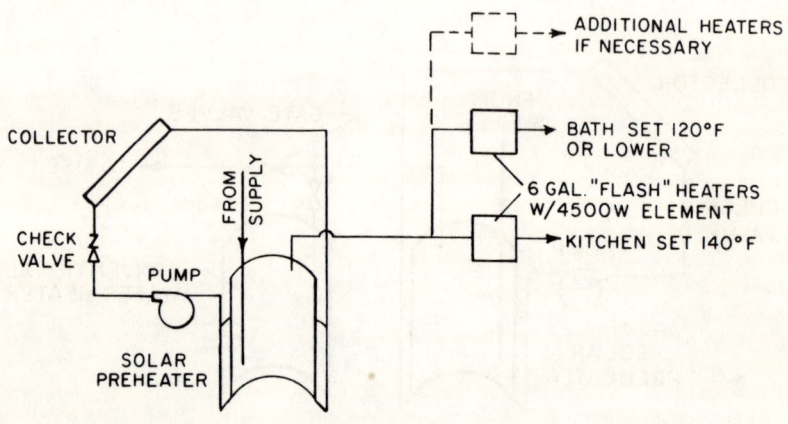

V-7. Schematic for Efficient Solar Heat Pre-Heat System

Other Storage Strategies. Even if one continues to consider thermal storage only in the form of sensible heat, there are yet other strategies which, in certain circumstances, may be successfully employed. One scheme which has been used with heat pumps is a deep well as a heat sink. In some instances, two wells—one "hot" and the other "cold"—have been employed for this type of application.

An application which piques the imagination is the use of heat pipes of a size comparable to those used to prevent melting of the permafrost along the Alaska pipeline. Use of these heat pipes would allow the use of natural thermal storage or the transfer of heat between ambient air, constant temperature earth, and storage.

Perhaps one of the most intriguing thermal storage schemes, particularly from an architectural viewpoint, is to totally or partially bury a building. In this manner the structure is not only well insulated, but the constant temperature of the surrounding earth, with proper design, may be used as infinite storage.

Of one matter we may be certain. As more solar energy systems are installed, and as more innovators apply their genius to the design of these systems, we shall surely witness a continuing variety of thermal storage materials, devices, and configurations.

REFERENCES

1. Shelton, Jay W. *The Woodburner's Encyclopedia.* Waitsfield, Vt.: Vermont Crossroads Press, 1976. Chapter One of this book contains an interesting history of the use of wood as a fuel.
2. Esbensen, T. V., and Korsgaard, V. "Dimensioning of the Solar Heating System in the Zero Energy House in Denmark," *Solar Energy,* Vol. 19, pp. 195–199.
3. Anderson, Bruce. *Solar Energy: Fundamentals in Building Design.* New York: McGraw-Hill, 1977, p. 228.
4. Dean, T. S. "Electric Analog Simulation for Geometry of Doubly Curved Tension Structures," *A.I.A. Research News,* Vol. II, May 1965. Numerical solutions of partial differential equations of the form $\nabla^2 \Phi = f(t)$ are frequently obtained by expanding the independent variable as an algebraic or trigonometric polynomial which can be fit to the boundary conditions. The cited reference solves the electric analog using a conductive paper (Teledeltos) having a known resistance.
5. Duffie, J. A., and Beckman, W. A. *Solar Energy Thermal Processes.* New York: Wiley Interscience, 1974, pp. 223–226. This reference also presents a model for stratification in pebble storage.
6. Close, D. J. "The Performance of Solar Hot Water Heaters with Natural Convection," *Solar Energy,* 6 (1962), pp. 33–40.
7. Pickering, E. E. "Residential Hot Water Solar Energy Storage," Proceedings of Solar Energy Storage Subsystems for the Heating and Cooling of Buildings, ASHRAE, 1975, 24–37. This reference reports an exhaustive study of containers for hot water.
8. Löf, G.O.G., and Hawley, R. W. "Unsteady State Heat Transfer Between Air and Loose Solids," *Industrial and Engineering Chemistry,* 40 (1948), pp. 1061–1070.
9. Hill, J. E. "Proposed Method of Testing for Rating Thermal Storage Devices Based on Thermal Performance," Proceedings of Solar Energy Storage Subsystems for the Heating and Cooling of Buildings, ASHRAE, 1975, pp. 101–105.
10. "Method of Testing Thermal Storage Devices Based on Thermal Performance," ASHRAE Standard 94-77, Feb. 1977.
11. Telkes, M. "Solar Energy Storage," *ASHRAE Journal,* Sept. 1974, pp. 38–44.
12. Telkes, M. "Storing Solar Heat in Chemicals—A Report on the Dover House," *Heating and Ventilating,* Reference Section, Nov. 1949, pp. 80–86.
13. Trombe, F., Robert, J. F., Cabanat, M., Sesalis, B. "Concrete Walls to Collect and Hold Heat," *Solar Age,* Aug. 1977, p. 13ff.
14. Niles, Philip W. B. "Thermal Evaluation of a House Using a Movable-Insulation Heating and Cooling System,." *Solar Energy,* Vol. 18 (1976), pp. 413–419.
15. Dean, T. S. "Tom and Jan Dean's House,." *Solar Age,* May 1977, pp. 22–25. See especially "The Specifics of the System," p. 25.
16. Balcomb, J. D., Hedstrom, J. C., and McFarland, R. D. "Simulation Analysis of Passive Solar Heated Buildings—Preliminary Results," *Solar Energy,* Vol. 19, pp. 277–282.
17. Balcomb, J. D., Hedstrom, J. C., and McFarland, R. D. "Thermal Storage Walls for Passive Solar Heating Evaluated," *Solar Age,* Aug. 1977, pp. 20–23.
18. Tabor, H. "Large-Area Solar Collectors for Power Production," *Solar Energy,* Vol. 7, 1963, pp. 189–194.
19. Tabor, H., and Matz, R. "Solar Pond Project," *Solar Energy,* Vol. 9, 1965, pp. 177–182.
20. Weinberger, H. "The Physics of the Solar Pond," *Solar Energy,* Vol. 8, 1964, pp. 45–56.

References

21. Rabl, A., and Nielsen, C. E. "Solar Ponds for Space Heating," *Solar Energy,* Vol. 17 (1975), pp. 1–12.
22. Styris, D. L., and Harling, O. K. "The Nonconvecting Pond Applied to Building and Process Heating," *Solar Energy,* Vol. 18 (1976), pp. 245–251.

TES Technologies Using Electrical Energy as the Primary Source

23. Biehl, R. A. "The Annual Cycle Energy System: A Hybrid Heat Pump Cycle," *ASHRAE Journal,* July 1977, pp. 20–24.
24. Cook, R. E., and Krubsack, R. M. "Energy Saving and Electric Load Balancing Potentials of Ice Bank Storage with Residential Air Conditioning," Paper No. CH-77-10-#3, *ASHRAE Transactions, 83* (1), 1977.
25. Hise, E. C., Moyers, J. C., Fischer, H. C. *Annual Cycle Energy System—Demonstration House Design Report.* Springfield, Va.: National Technical Information Service, 1976.

Low Temperature TES Technology Assessments

26. Asbury, J. G., and Mueller, R. O. *Solar Energy and Electric Utilities: Can They Be Interfaced?* Springfield, Va.: National Technical Information Service.
27. Asbury, J. G., and Mueller, R. O. *Science,* 195, 4 Feb. 1977, pp. 445–450.
28. Asbury, J. G., et al. *Assessment of Energy Storage Technologies and Systems.* Springfield, Va.: National Technical Information Service.
29. Golibersuch, D. C., et al. *Cool Storage Assessment Study.* Palo Alto, Ca.: Electric Power Research Institute, 1977.

APPENDIX
Conversion Factors: Common Relationships

Multiply:	by:	to obtain:
Acres	43,560	square feet
Acre-ft.	1,233.5	cubic meters
Angstrom units	1×10^{-8}	centimeters
Barrels, oil (crude)	5,800,000	Btu (energy)
Barrels, oil	5.615	cubic feet
Barrels, oil	42	gallons
Board feet	0.0833	cubic feet
Btu	0.55556	chu (centigrade heat units)
Btu	777.48	foot-pounds
Btu	1,055	joules
Btu	0.29305	watt-hours
Btu/hr/ft^2/°F	5.682×10^4	watts/cm^2/°C
Btu per square foot	0.271	langleys (calories per sq. cent.)
Btu/hr/ft^2 (°F/in)	1	chu/hr/ft^2 (°C/in)
Calories	3.9685×10^{-3}	Btu
Calories	4.184	joules
Centigrade heat units (chu)	1.8	Btu
Common brick, number of	5.4	pounds
Cords 128		cubic feet
Cubic feet	0.037037	cubic yards
Cubic feet	7.48	gallons
Cubic feet of water	62.37	pounds at 60°F
Cubic feet of common brick	120	pounds
Cubic feet per second	448.83	gallons per minute
Cubic yards of sand	2,700	pounds
Feet of water (39.2°F)	0.4335	pounds per square inch
Feet of water	0.88265	inches of mercury at 0°C
Gallons	0.1337	cubic feet
Gallons of water	8.3453	pounds of water at 60°F
Horsepower	33,000	foot-pounds per minute
Horsepower	42.42	Btu per minute
Horsepower	2,546	Btu per hour
Horsepower	1.014	metric horsepower
Horsepower	0.7457	kilowatts
Horsepower, metric (chevalvapours)	0.9863	horsepower
Inches of mercury at 32°F	0.4912	pounds per square inch
Kilowatts	56.90	Btu per minute
Kilowatts	1.341	horsepower
Kilowatt-hours	3,413	Btu
Kilowatt-hours	2.66×10^6	foot-pounds
Langleys (cal/cm^2)	3.69	Btu per square foot
Langleys per minute	0.0698	watts per square centimeter
Microns	1×10^{-4}	centimeters
Months (mean calendar)	730.1	hours
Newtons	0.22481	pounds (force)
Pounds of water	0.1198	gallons
Pounds of water evaporated at 212°F	970.3	Btu
Pounds per square inch	0.068046	standard atmospheres
Pounds per square inch	51.715	millimeters of Hg at 0°C
Standard atmosphere	14.696	pounds per square inch
Tons (long)	2,240	pounds
Tons (short)	2,000	pounds
Tons (short)	0.907185	metric tons
Tons (metric)	2,204.62	pounds
Tons of refrigeration	12,000	Btu per hour
Therms	1×10^5	Btu
Watts	3.413	Btu per hour
Watts	0.00134	horsepower

(From Anderson, p. 329)